Home Gardens for Improved Food Security and Livelihoods

T0133756

Home Gardens for Improved Food Security and Livelihoods demonstrates how home gardens hold particular significance for resource-poor and marginalized communities in developing countries, and how they offer a versatile strategy toward building local and more resilient food systems.

With food and nutritional security being a major global challenge, there is an urgent need to find innovative ways to increase food production and diversify food sources while increasing income-generating opportunities for communities faced with hunger and poverty. This book shows that when implemented properly, home gardens can become just such an innovative solution, as well as an integral part of sustainable food security programs. It provides a conceptual overview of social, economic, environmental and nutritional issues related to home gardening in diverse contexts, including gender issues and biodiversity conservation, and presents case studies from Africa, Asia and Latin America highlighting home gardening experiences and initiatives. The volume concludes with a synthesis of key lessons learned and ways forward for further enhancing home gardens for sustainable food security and development.

This book will be a useful read for students and scholars working on local food systems, food security, sustainable development and more broadly development strategy.

D. Hashini Galhena Dissanayake is Assistant Professor in the College of Agriculture and Natural Resources at James Madison College.

Karimbhai M. Maredia is Professor in the College of Agriculture and Natural Resources, Michigan State University.

Other books in the Earthscan Food and Agriculture Series

For more information about this series, please visit: www.routledge.com/books/series/ECEFA/

Home Gardens for Improved Food Security and Livelihoods

Edited by D. Hashini Galhena
Dissanayake and
Karimbhai M. Maredia

Routledge
Taylor & Francis Group

LONDON AND NEW YORK

First published 2021
by Routledge
2 Park Square, Milton Park, Abingdon, Oxon OX14 4RN

and by Routledge
52 Vanderbilt Avenue, New York, NY 10017

Routledge is an imprint of the Taylor & Francis Group, an informa business

British Library Cataloguing-in-Publication Data
A catalogue record for this book is available from the British Library

Library of Congress Cataloging-in-Publication Data
A catalog record for this book has been requested

ISBN: 978-1-138-20213-9 (hbk)
ISBN: 978-0-367-50296-6 (pbk)
ISBN: 978-1-315-47177-8 (ebk)

Typeset in Bembo
by Apex CoVantage, LLC

This book is dedicated to scholars and practitioners who have dedicated their lives to advancing and sharing knowledge on local food systems.

Contents

Preface and Acknowledgement

Feeding the estimated population of nine billion people in 2050 is a monumental task, and every contribution to the food supply will matter and will help the fight against hunger and malnutrition. Cultivation of small garden plots adjacent to human settlements is one of the oldest agricultural practices found all over the world. In addition to standing the test of time, evidence suggests that home gardening provides a variety of socioeconomic and environmental benefits, particularly to the home gardeners themselves, their families and their communities. Small but important contributions from home gardens to human well-being and ecological services in the aggregate enhance communal, regional and national food sovereignty and prosperity. As such, home gardening is widely used as a developmental strategy and intervention. Home gardening has proven to be a useful solution, particularly in situations where resources are limited, institutions are weak and rapid interventions are needed to address hunger, malnutrition and poverty.

Home gardens can also play an important role in empowering women in their communities. While most outcomes, characteristics and functions of home gardens remain convergent and universal, there are some unique qualities and differential experiences from home gardening. Our motivation for this book lies in our interest to share aspects and outcomes of home gardening which are common and illustrate experiences that are unique through case studies from various parts of the world. We hope that while adding to scholarship on the subject, sharing of these divergent perspectives and experiences will focus attention on home gardens as a versatile strategy for enhancing household food security, nutrition and income.

This book brings together the perspectives and experiences from home gardening projects and programs of scholars and practitioners from diverse disciplines. The book was inspired by a doctoral research project conducted by us in conflict-affected areas in northern Sri Lanka. After decades of conflict, civil war in Sri Lanka ended in 2009. The Sri Lankan government at the time initiated home gardening programs to assist the displaced and resettled families in the conflict-affected areas in the north and east that were also seriously affected by the 2004 tsunami.

Our research provided new insights on the role of home gardens as a potential post-disaster strategy. In addition to the experiences from Sri Lanka featured in Chapter 8, this book consists of several chapters presenting interesting case studies that highlight the utility of home gardens in various contexts. Some authors delve into identifying constraints, areas for improvement and potential for scale-up to make home gardening attractive and sustainable.

Chapter 1 contains a literature review consisting of definitions, concepts and themes providing the theoretical base to the discussion on home gardens. The next three chapters will analyze more specifically contributions of home gardens to nutritional security (Chapter 2), biodiversity conservation (Chapter 3), and gender dimensions (Chapter 4). Chapters 5 and 6 through 10 illustrate case studies of various applications of home gardens from around the world.

We would like to thank all the collaborating authors for their scholarly contributions. We hope this book deepens your understanding of home gardens and also provides novel scholarly perspectives and insights.

D. Hashini Galhena Dissanayake
Michigan State University, USA

Tables, figures, boxes and photos

Tables

Figures

Boxes

Photos

Contributors

Adrienne Attorp, University of Reading, United Kingdom

Helga Blanco-Metzler, Sede Guanacaste, University of Costa Rica

Nidhi P. Chanana, The Energy and Resources Institute, India

Neetika W. Chhabra, The Energy and Resources Institute, India

Julien Curaba, The World Vegetable Center (AVRDC), Taiwan

Jacqueline d'Arros Hughes, The World Vegetable Center (AVRDC), Taiwan

Peter Dorward, University of Reading, United Kingdom

Alex Diaz Porras, Ministry of Agriculture, Montes de Oro, Costa Rica

Russell Freed, Michigan State University, USA

D. Hashini Galhena Dissanayake, Michigan State University, USA

Ramjee Ghimire, Michigan State University, USA

Joseph Guenthner, Idaho State University, USA

Danny Hunter, Bioversity International, Italy

Katarina Huss, Michigan State University, USA

Nanda Joshi, Michigan State University, USA

Amir Kassam, University of Reading, United Kingdom

John Donough Heber Keatinge, The World Vegetable Center (AVRDC), Taiwan

Karimbhai M. Maredia, Michigan State University, USA

Gunasingham Mikunthan, University of Jaffna, Sri Lanka

Kamlesh Panchole, Aga Khan Rural Support Programme, India

Naveen K. Patidar, Aga Khan Rural Support Programme, India

Sajan Prajapati, Aga Khan Rural Support Programme, India

Gamini Pushpakumara, University of Peradeniya, Sri Lanka

Linda Racioppi, Michigan State University, USA

Jessica Sokolow, Cornell University, USA

Bhuwon Sthapit, Bioversity International, Italy

Wawan Sujarwo, Indonesian Institute of Sciences, Indonesia

Foreword

Food security continues to be a major challenge for millions of people around the world. Global challenges include climate change, demand for water, ecological disruptions, population growth, hidden hunger, economic instability and food security. The projected world population of 9.5 billion people by 2050 will only accelerate these challenges. Home gardens can play an important role in minimizing our current challenges.

This book outlines how home gardens can be a very effective way to quickly enhance household food security, nutritional needs and economic well-being. Home gardens will give a jump-start to household economic growth and enhance nutritional security for families. The production of fruits, vegetables, herbs, spices, ornamentals, medicinal and plantation crops and livestock in home gardens is an eco-friendly practice to improve food and nutritional security. Consumption of a variety of fresh fruits and vegetables helps to improve family health and nutrition, which is especially important for young children. Home gardens are also an effective tool in empowering women with added income. Especially for resource-poor communities, home gardens are an excellent first step for establishing local food systems that improve economic growth and enhance household nutrition.

Chapter 1 gives an overview of the relevant literature on home gardens and shows how home gardens impact the food security and nutrition of households. Chapter 2 reviews how home gardens can enhance nutritional security for families. Chapter 3 reviews the issues relating to the conservation of biodiversity. Chapter 4 reviews the role of gender and home gardens. Chapter 5 gives a case study of nutritional health in Mozambique. Chapters 6 and 7 examine home gardens in Costa Rica and India. Chapter 8 outlines how home gardens can provide food security in post-crisis Sri Lanka. Chapter 9 reviews how livestock can be incorporated into home gardens. Chapter 10 gives an overview of a home garden program in the poorest of rural India. Chapter 11 concludes with lessons learned and ways to enhance home garden programs.

This book shows how home gardens can make a significant impact in improving food security, enhance economic growth and improve the nutritional status of families. It also is a good addition to home garden literature. Congratulations to D. Hashini Galhena Dissanayake and Karimbhai M. Maredia for putting this important book together.

Russell Freed
Professor Emeritus, Michigan State University, USA

1 Understanding the global practice of home gardening

D. Hashini Galhena Dissanayake

Home gardens have been an integral part of local farming and food systems since the beginning of civilization. The very beginning of modern agriculture dates back to subsistence production systems that began in small garden plots around the household. In fact, domestication and conservation of crops, especially horticultural crops, began in home or kitchen gardens, hence the word "horticulture" is derived from the Latin words *hortus* meaning garden and *cultūra* meaning cultivation. In establishing gardens, ancient Egyptians were attentive to the symbolical meaning and form (Wilkinson, 1994). An assortment of vegetable, fruit and root crops, flowers and vines were selected and planted accordingly (Janick, 2002). The Mayans discovered and perfected the practice of multi-cropping in their gardens (von Baeyer, 2010). The Greeks and Romans maintained kitchen gardens next to their households. The Greeks applied gardening knowledge and practices of the Persians and fine-tuned gardening techniques such as enclosed growing, grafting, budding and rotation. Across the world, home gardens have persistently endured the test of time and continued to play an important role in providing food and income for families (Marsh, 1998).

Defining home gardening

Home gardening is practiced widely all over the world. In literature, home gardens are referred to as kitchen, backyard, dooryard, household, homestead and mixed gardens as well as farmyard enterprise and compound cultivation (Niñez, 1987; Fresco and Westphal, 1988; Midmore et al., 1991). Many definitions of home gardens can be found in literature derived through spatial observation and field research. G.J.A. Terra pioneered the literature on home gardens in the early 1950s with his publications on "Mixed Garden Horticulture in Java" and defines the home garden as "an area of land, individually owned, surrounding a house and usually planted with a mixture of perennials and annuals" (Terra, 1954).

Home gardens are predominantly small-scale subsistence agricultural system and are found across the rural, suburban, and urban landscape (Niñez, 1987; Nair, 1993). Vera Niñez (1987) composes a more elaborate definition that attempts differentiate home gardens from other agricultural production systems and states,

> "the household garden is a small-scale production system supplying plant and animal consumption and utilitarian items either not obtainable, affordable, or readily available through retail markets, filed cultivation, hunting, gathering, fishing, and wage earning. Household gardens tend to be located close to dwelling for security, convenience, and special care. They occupy land marginal to field production and labor marginal to major household economic activities. Featuring ecologically adapted and complementary species, household gardens are marked by low capital input and simple technology." Another scholar writes that home gardens are "cropping systems characterized by the permanent use of small, mostly fenced, plots at a short distance from the homestead with a variable number of annual, biennial and/or perennial intercrops giving rise to a multi-story physiognomy."

> (Fresco and Westphal, 1988, p. 405)

Instead of providing any definition or accepting those of others, some authors merely describe home gardens as "intimate, multi-story combinations of various trees and crops, sometimes in association with domestic animals, around homesteads" (Kumar and Nair, 2004, p. 135). Some authors identify home gardens as an integral part of the local land-use system and the watershed (Fresco and Westphal, 1988) and part of the larger agricultural landscape: "home gardens are microenvironments within a larger farming system that contain high levels of species diversity and may contain crop species or varieties of species different from those found in surrounding agroecosystems" (Eyzaguirre and Linares, 2004). Other definitions illustrate the home garden as a parcel of land near the family dwelling that spreads vertically and horizontally that serves multiple purposes, holding multiple species of crops and animals (Photo 1.1). For instance, Hoogerbrugge and Fresco (1993) state that "a home garden is a small-scale, supplementary food production system by and for household members that mimics the natural, multi-layered ecosystem." This refers to a perpetual and subsistent food system with the primary goal to fulfill and supplement household needs.

The diverse plants and animal products from home gardens have multiple uses. For instance, herbs and weeds are often used in indigenous medicines and teas and as aromatics, seasonings, dyes, green vegetables and fibers for weaving (Buchanan, 2012; Gökçebağ and Özden, 2017). The livestock in the home garden provides a source of protein-rich food, alternative fuel and energy, and fertilizer and manure for composting (Pulami and Paudel, 2004). Compost containing livestock, kitchen and yard waste is often the primary source of nitrogen (N), phosphorus (P), potassium (K) and organic matter for the home garden (Photo 1.2). Though output from the home gardens is mainly intended for family consumption, the surplus output from the home garden can be sold

Photo 1.1 A typical home garden in the humid tropics
Source: D. H. Galhena Dissanayake.

for additional income (Mendez et al., 2001; Yiridoe and Anchirinah, 2005; Sthapit et al., 2006; Ferdous et al., 2016). In addition to these obvious outcomes, home gardens provide several ecological benefits, from air purification to preservation (Galluzzi et al., 2010; Calvet-Mir et al., 2016). As such, there is much complexity to what the home gardens can do, which perhaps explains the lack of consensus among scholars on a universal definition.

Factors that characterize home gardens

The body of literature defining home gardens has continued to expand extensively to integrate characteristic variations across space, species diversity and composition, range of functions, cultural values, social and economic impacts,

Photo 1.2 Composting pile in home garden
Source: D. H. Galhena Dissanayake.

conservation and environmental aspects, adding various dimensions and perspectives to the discourse (Kumar and Nair, 2004; Galluzzi et al., 2010; Pushpakumara et al., 2012; Langellotto, 2014; Igwe et al., 2014). There is also a
growing body of literature showcasing the gender-related issues surrounding
home gardens (Zypchyn, 2012; Schreinemachers et al., 2015; Nguyen et al.,
2017). However, as Gupta (1989) pointed out, literature on home gardens at
times may not adequately reflect the values and opinions of the home gardener,
as the background, gender and predispositions of the scholars take precedence
and influence the way various arguments around home gardens are presented.
Given this background, attempts to characterize home gardens include some
biases resulting from limitations in the research design, sampling and interview
process, and reporting. Mitchell and Hanstad (2004), referring to research by

Brownrigg (1985) and Marsh (1998), provide five attributes to a home garden: (1) it is located near the residence; (2) it occupies a small area; (3) it contains a high diversity of plants; (4) it provides a supplemental source of food and other materials for family consumption and income generation; and (5) it is a practice that the poor can easily adopt. Moreover, they are described to be low input and low management production systems (Huai and Hamilton, 2009; Gautam et al., 2009).

Home gardens are generally viewed as agroforestry units, small-scale food systems or both (Torquebiau, 1992; Maroyi, 2009; Pushpakumara et al., 2012). They are delimited by physical demarcations such as live fences or hedges, fences, ditches or boundaries established through mutual understanding. Based on the ecology, home gardens may be of two categories: tropical or temperate (Niñez, 1985). There is a plethora of research and case studies focusing on tropical home gardens, primarily in Central and South America (Del Angel-Pérez and Mendoza, 2004; Albuquerque et al., 2005; Perrault-Archambault and Coomes, 2008; Vazquez-Garcia, 2008; Aguilar-Støen et al., 2009). There is also a growing body of literature highlighting various dimensions of home gardens from Asia (Perera and Rajapakse, 1991; Talukder et al., 2000, 2006; Abdoellah et al., 2001; Cai et al., 2004; Ali, 2005; Sunwar et al., 2006) and Africa (Tchatat et al., 1996; Drescher et al., 1999; High and Shackleton, 2000; Faber et al., 2002; Faber and Wenhold, 2007; Maroyi, 2009). Home gardens in temperate areas have also been discussed widely (Cleveland et al., 1985; Seeth et al., 1998; Agelet et al., 2000; Thompson et al., 2003; Morton et al., 2008; Jesch, 2009; Rigat et al., 2009; Aceituno-Mata, 2010; Bassullu and Tolunay, 2010; Reyes-García et al., 2010; Calvet-Mir et al., 2012). Several studies focus on home gardens in industrialized countries (Cleveland et al., 1985; Bleasdale et al., 2011). Depending on the location, home gardening is practiced year-round or seasonally.

Several articles note that home gardens share some similarities across space, and yet they maintain some differentiation with regards to structure, functionality and composition (Fernandes and Nair, 1986; Soemarwoto and Conway, 1991; Torquebiau, 1992). In addition to the natural ecology of the location, such variation is due to factors such as family assets base (land, capital, etc.), inputs access (seeds, labor, etc.), know-how, enthusiasm and preferences (Christanty et al., 1986; Asfaw, 2002). Wiersum (2006), through observation of home gardens in Indonesia, showed that the structure, composition, intensity of cultivation and diversity of the gardens changed with the socioeconomic status of the household. As families became economically stable, their choice of crops changed from staples to horticultural crops, and some expanded to integrate livestock production (Photo 1.3). Therefore, household economics is a significant determinant of the characteristics of home gardens. Niñez (1985) classified home gardens into four types: survival gardens, subsistence gardens, market gardens and budget gardens.

The divergence among home gardens also depends on the availability of land, and it is a key determinant of the form and organization of the home

Photo 1.3 Cash crops grown in home gardens
Source: D. H. Galhena Dissanayake.

garden. As such, the size of a home garden changes from household to household (Hoogerbrugge and Fresco, 1993). Often, if the families have access to land, a home garden is established near the homestead in land that is not intended for commercial cultivation because of its size, topography or location. Near landless and landless households have found creative ways to establish home gardens in their limited space (Rammohan et al., 2019). As such, home gardeners cultivate crops on dedicated or partially committed plots, vacant spaces on the outskirts of the property and in containers, and they may be organized into multiple vertical layers. The reuse and recycling of household and garden waste products such as coconut husks, tin cans, plastic bags, bottles and containers, PVC and rubber pipes, and even tires are common (Photos 1.4a and 1.4b). Limited availability of garden space, soil quality, and household economics and awareness factors have a bearing on the structure and organization of the garden.

Annual vegetables, fruits, herbs and other home garden products used on a day-to-day basis are commonly located nearest to the living quarters or kitchen. Perennial fruit, palm, spice, fodder and timber species are situated on the peripheral contours of the property. The home gardener's decision to retain the trees that came with the land, cut them down or plant new ones has effects on the composition and biodiversity of the garden as well as the benefits it provides. Depending on the home gardener's interest and value for aesthetics, the gardens will contain flowering and ornamental plants. Gardeners often cultivate flowering plants such as marigolds and chrysanthemums to control pests and provide for cultural purposes. While some medicinal plant species may voluntarily grow in home gardens, gardeners knowledgeable in medicinal properties of plants may actively protect or propagate them. Some home

Photo 1.4a Old tire used to cultivate vegetables

Photo 1.4b Leafy vegetables cultivated in plastic soda bottles
Source: D. H. Galhena Dissanayake.

gardeners diversify their home gardening activities to integrate livestock such as ruminants, poultry and fish (Photo 1.5).

Water availability and soil type and fertility are key factors determining the composition and functionality of the home garden (Trinh et al., 2003). It is important to ensure good drainage during the rainy season and irrigation during the dry season. Home gardeners rely on rain as well as ground and surface water available to them to irrigate their home gardens. Thus home garden production is vulnerable to changes in environmental conditions and adverse weather such as drought and floods. While home gardening activities necessitate some basic knowledge in horticulture and agronomics, access to information and training can help soften the negative impacts and losses from

Photo 1.5 Livestock in home garden
Source: D. H. Galhena Dissanayake.

such adversities. Soil type and health also determines the diversity and composition of the crop species (Kehlenbeck and Maass, 2004). Home gardens in
areas with adequate water availability throughout the year along with good
soil quality provide suitable conditions for plants to thrive, resulting in complex multi-strata home garden systems (Christanty et al., 1986).

The output from home gardens is primarily used for family consumption
and subsistence; any excess is shared, exchanged or sold. Depending on household needs, resource availability and environmental conditions, home gardening tends to be quite dynamic (Eyzaguirre and Linares, 2004; Sthapit et al.,
2004). The home garden is managed by the family members, and their time
and labor is critical for garden management (Sthapit et al., 2004). Women,

children and elders play an important role in their management (Fernandes and Nair, 1986; Niñez, 1987; Landauer and Brazil, 1990; Torquebiau, 1992; Jose and Shanmugaratnam, 1993; Wojtkowski, 1993; Dash and Misra, 2001). The hiring of wage laborers for cultivation and maintenance of the home garden is dependent on the economic capacity of the household. Analogous to Boserup's (1965) observations for agriculture, increase in labor availability leading to intensification of home gardening can directly affect the garden configuration. Availability of family labor and the household's consumption and income generation motives stimulate decisions that drive home gardening activities such as crop selection, input procurement, garden management, harvesting, processing, sale of products and so forth (Ali, 2005). In addition, the development and composition of home gardens are impacted by opportunities for off-farm employment, size and composition of the family and local customs (Moreno-Black et al., 1996).

Home gardens can be characterized as grounds for experimentation. Home gardeners test varieties, experiment with cultivation and management practices, and use their garden as a demonstration and dissemination site. The emergence of new innovations and techniques have helped improve home garden production and have even made it possible with little to no land (Ranasinghe, 2009). For instance, conversion of kitchen waste, animal manure and litter, and other organic residues into compost and vermicompost (Photos 1.6a and 1.6b) has helped to enhance soil fertility and in turn considerably increase productivity (Hoogerbrugge and Fresco, 1993).

Niñez (1987) uses 15 type-specific attributes to characterize home gardens. These attributes as they apply to home gardens are compared to those of commercial cropping systems in Table 1.1. This comparison helps to distinguish some inherent features of home garden cultivation from commercial agriculture.

Home gardens as a development strategy

Hunger and malnutrition are key development issues facing populations in developing countries (Food and agriculture organization of the United Nations (FAO), 2010), where they suffer from various forms of food insecurity.[1] The need to increase local food production and maintain buffer stocks will continue as the demand for food grows within countries with bulging populations and as a mechanism to mitigate the effects of global economic and political volatilities. According to the Food and Agriculture Organization (FAO 2009), by 2050 there will be more mouths to feed and food production will need to be increase by 70% in order to meet the average daily caloric requirement[2] of the world's population. The countries that have to put in the greatest effort are not surprisingly the developing countries. This emphasized the need for interventions and strategies to increase productivity amid resource and input constraints, the rising cost of production and changing climates. Thus one strategy alone will not suffice. It will require diverse interventions involving multiple stakeholders from the policy level to the grassroots level. Home gardens have endured

Photo 1.6a Low-cost and simple vermicompost production unit

Photo 1.6b Vermicompost

Source: D. H. Galhena Dissanayake.

Table 1.1 Comparison of home gardens to commercial cropping systems

Characteristic	Home gardens	Commercial cropping systems
Species density	High	Low
Species type	Staples, vegetables, fruits, medicinal plants	Specialized crop production (may involve some intercropping)
Production objective	Home consumption	Commercial purposes
Labor source	Family labor (men, women, elderly and children)	Farmer's own labor and/or hired labor
Labor requirements	Part-time	Full-time or part-time depending on the time of the season
Harvest frequency	Daily, seasonal	At the end of the growing season
Space utilization	Horizontal and vertical	Mostly horizontal (depending on the crop, vertical cultivation may be used)
Location	Near dwelling	Varies with access and availability of land
Cropping pattern	Irregular, sometime structured	Organized
Technology	Simple hand tools	Simple implements to complex machinery
Input cost	Low	Medium to high
Distribution	Rural and urban areas	Mostly rural and peri-urban
Skills	Gardening and horticultural skills	Good farming and management skills
Assistance	None or minor	Yes

Source: Adapted from Niñez (1987).

the test of time and are viewed as efficient and sustainable land-use systems of agroforestry and food production (Torquebiau, 1992). They can be established in various agroecological zones and can be adopted and practiced by especially marginalized groups with little or no access to resources and institutional support. Further, home gardening can be tailored to fit the cultural norms and traditions of the household, which is necessary for any intervention to gain acceptance and to sustain (Bandarin et al., 2011). Hence, home gardens have seamlessly integrated into agriculture and food production systems in many developing countries and are widely used as an intervention to alleviate hunger and malnutrition (Johnson-Welch et al., 2000).

Benefits of home gardens

Exploration of past and present literature on home gardens unveils a number of sociocultural and economic impacts to households. Some common benefits are discussed in this section.

Increase food availability at household level

Home gardens provide a broad base of edible products which include some cereals, vegetables, fruits, herbs and spices as well as animal products from livestock raised in the garden (Karyono, 1990; Michon and Mary, 1994;

Albuquerque, 2005; Angel-Pérez and Mendoza, 2004; Kumar and Nair, 2004; Peyre et al., 2006). A study shows that Pacific Islanders obtained their staple root crops such as taro, cassava, tannia or cocoyam, sweet potato, greater yam and sweet yam from their home gardens (Thaman, 1995). Studies from South Asia, Southeast Asia, South America and Africa also confirms these findings (Bennett-Lartey et al., 2001; Coomes and Ban, 2004; Krishna, 2006; Kehlenbeck et al., 2007; de la Cerda and Mukul, 2008; Maroyi, 2009). Wiersum (2006) states that home gardens are especially beneficial to resource-poor families than well-to-do families with access to land and capital as they make available staples and secondary staples to the households. Marsh (1998) reports that households with home gardens can fulfill more than half of their household requirement of vegetables, fruits, tubers and yam. Food availability to those with limited access to arable land (Photo 1.7), living in heavily degraded or densely populated areas can be increased using home gardens (Soemarwoto

Photo 1.7 Containers made from Palmyra palm used for vegetable production in sandy areas
Source: Gunasingham Mikunthan

and Conway, 1991; Abebe et al., 2006). In addition to people, home gardens also provide foodstuffs and fodder for the animals raised by the family to obtain animal-derived food products.

With their proximity to the family dwelling, home gardens can provide easy access to plant and animal food products. The products derived from the gardens help enhance energy and nutritional requirements of the family. A study by Ochse and Terra (1934) is one of the early studies investigating this benefit of home gardening. They found that home gardens of households in Central Java, Indonesia, contributed to 18% of the caloric and 14% of the protein intake. In a subsequent study, Ochse (1937) link household nutritional status to the development of home gardens. Danoesastro (1980) saw a rise in household food availability and consumption with the intensification of home garden production. Other studies from around the world have also reported similar observations (Niñez, 1985; Brownrigg, 1985; Ferdous et al., 2016).

Improve physical, economic and social access to food

Plants and livestock grown and raised in home gardens provide physical access to a diverse array of fresh foods rich in nutrients (Marsh, 1998). The family can then prepare or process these foods according to their preferences and consume them immediately or later. The findings from Bangladesh revealed that products from livestock in home gardens were frequently the only source of animal protein accessible to the poor families (Marsh, 1998). Wiersum (2006), reflecting on home gardens in Java and Sulawesi in Indonesia, notes that they make available a small but a steady flow of food products for household consumption, though the type of food products obtained from the garden often depends on environmental, socioeconomic and cultural factors (Peyre et al., 2006; Abdoellah et al., 2006). Home gardens can help connect families who are isolated by geographic barriers or having limited access to markets to homegrown food, enhancing their food security and well-being.

Improving economic access to food is another advantage of home gardening. A known but not well-documented aspect of home gardening is the savings it provides to the household by reducing its expenditure on food (Langellotto, 2014). Home gardening is a low-cost activity that primarily employs family labor. Based on ten sources that assessed the economic costs and yields of vegetable gardens, Langellotto (2014) concludes that if the fair market value of labor was excluded from the costs, vegetable gardens returned profits. The reduced food bill enables households to divert funds toward the purchase of other food items or set them aside for other expenses.

Social factors such as cultural norms and entitlements of individuals result in differential access to food within households. Populations vulnerable to food insecurity include, among others, poor individuals and families, ethnic minorities, refugees, women, children and individuals with disabilities. Home gardens have proven the possibility of reducing social barriers to food security by being a pro-poor (Ferdous et al., 2016), post-disaster solution (Galhena et al., 2012) and social empowerment strategy (Patalagsa et al., 2015). Most social systems tend to prioritize men. They control many of the household's resources including land and income. Women are often the primary caretakers of home gardens. Home gardening is a favorable approach to empower women in patriarchal societies, as it is perceived a part of household chores and it does not challenge men's authority in the household. A home gardening and nutrition education project in Bangladesh revealed that women had greater opportunities to exercise control and make decisions over home gardening activities and the allocation of outputs and income generated from garden sales (Marsh, 1998).

Diversify diets

Undernourishment and nutrient deficiency are major public health issues. Nutrient deficiencies resulting from poor access to proteins, vitamins and minerals are responsible for escalating the susceptibility to infectious diseases, impairing physical and mental development, and increasing the risk of child and maternal mortality in the developing countries. Nearly 6% of deaths of children under five years of age is attributed to vitamin A deficiency (Stoltzfus et al., 2004). Output from home gardens can supplement the staple-based diet by adding nutrient-rich food items that are rich in macro- and micronutrients (Kumar and Nair, 2004; Wiersum, 2006; Soemarwoto, 1987; Hoogerbrugge and Fresco, 1993; Abdoellah et al., 2001). A study conducted in KwaZulu-Natal in South Africa found that production of leafy vegetable in home gardens significantly improved vitamin A availability in young children (Faber et al., 2002). The national home gardening program in Bangladesh, supported by Helen Keller International together with local non-governmental organizations, was also successful in increasing the availability and consumption of foods rich in vitamin A (Talukder et al., 2000; Helen Keller International-Asia-Pacific, 2010). Improvements to crop diversity can greatly contribute to dietary diversity (Rammohan et al., 2019).

Human health is directly linked healthy food habits and lifestyle. In addition to whole foods and livestock products, home gardens contain a diversity of plants that can be used as herbs and for medicinal purposes. In fact, medicinal plants were identified to be the second most important group of plants next to cash crops (Perera and Rajapakse, 1991; Millat-e-Mustafa et al., 2002). Of the 125 plant species found in Kandyan gardens in Sri Lanka, 30% were exclusive used for medicinal purposes. In Catalina in Italy, home gardens harbored close to 250 medicinal plant species (Agelet et al., 2000).

Increase household's resilience to food shortages

There is growing body of literature to support the fact that home gardens can be a thrifty solution to address food and nutritional insecurity in challenging situations. As such, home gardening projects have been the focus of policy and developmental strategies. Innovations have made it possible for even the poorest, resource-poor and landless households to establish and maintain home gardens with a few inputs. Buchmann (2009) notes that home gardens in Cuba were used as a resilience strategy to sustain food security in the face of economic crisis and political isolation. The gardens helped to supplement the basic food staples received as rations (Wezel and Bender, 2003) and have been instrumental in reducing "hidden hunger" and micronutrient deficiency in the country.

Home gardens can support individuals and families facing human-made and natural disasters, de-modernization and various other internal and external shocks to mitigate negative effects on food security. Tajikistan gained independence from the Soviet Union in 1991, and with its agricultural production almost exclusively turned to cotton production, the country faced many challenges to achieving food security in the years that followed. Nevertheless, William Rowe (2009) found that kitchen gardens played a key role in curtailing food insecurity and was instrumental in providing as much as one-third of the food products sold in the local markets.

Home gardens have been proposed as a strategy to address hunger, malnutrition and poverty to overcome adversities resulting from natural disaster and conflict (Marsh, 1998; Wanasundera, 2006; Ibnouf, 2009; Iannotti et al., 2009; Kirabo et al., 2011; Keatinge et al., 2012). Political instability has escalated hunger and poverty in countries around the world (von Grebmer et al., 2011), and eradicating hunger is a key precondition to peace-building. Thus projects focusing on home gardens offer a culturally appropriate, low-cost solution that can be easily replicated.

Home gardens are a robust system of cultivation (Photo 1.8) that can withstand numerous resource limitations (Salam et al., 1995; Holden et al., 1996; Ranasinghe, 2009). For families living in slum areas in the Peruvian capital of Lima, home gardening increased the availability and access to a variety of staples, vegetables and fruits that would have otherwise being inaccessible to them due to poverty (Niñez, 1985). In Ethiopia, enset-coffee home gardens not only provided subsistence and complementary food products and income generation opportunities, but they also elevated the household ability to cope with food shortages during famines (Abebe et al., 2006).

Providing supplementary income generation and employment opportunities

The benefits of home gardening go beyond food and nutritional security. In the absence of employment prospects, home gardening can provide revenue options and serve a supplemental source of income. Evidence shows economic incentives from home gardening resulting from income generation, livelihoods

Photo 1.8 Subsistence home garden
Source: D. H. Galhena Dissanayake.

and entrepreneurship opportunities (Mendez et al., 2001; Trinh et al., 2003; Calvet-Mir et al., 2012). More than a primary means of income generation, home gardens are often considered a source of supplemental income through the sale of various products. Trinh (2003) reports that families in mountain areas of Vietnam generated on average 22% of their income from home gardening, and Okigbo (1990) finds the value to be more than 60%. Income from the sale of fruits, vegetables and livestock products from home gardens allowed households to use the proceeds to purchase other food items, increase their savings and spend on other needs (Vasey, 1985; Iannotti et al., 2009). Furthermore, through value addition, home garden products can fetch a higher price and lead to home-based enterprises.

Delivering environmental benefits

Home gardening often involves ecologically friendly approaches to food production with minimal impact on the environment. As discussed by Mitchell and Hanstad (2004), rich species diversity and composition of flora and fauna are a key features of home gardens. Bompard et al. (1980), through their investigation of home gardens in Indonesia, found 240 (sub-)species reported for a single garden. Similarly, Kehlenbeck and Maass (2004) identified 149 plant species in the 30 home gardens they assessed in Sulawesi, Indonesia. The diversity in home gardens also contributes to in situ conservation and preservation of germplasm (Santhoshkumar and Ichikawa, 2010). With the expansion of comercial agriculture, some landraces and spices have been confined to home gardens. Therefore, home gardens act as a living repository for genetic materials and divrsity.

Home gardens provide a number of ecosystem services. Calvet-Mir et al. (2016) divide ecosystem services provided by home gardens into four categories – regulating, habitat, production, and cultural services. The ability of home garden to avert floods, maintain soil quality, enhance water quality and act as a defense against pests and disease fall under its regulating services. Habitat services result from the natural environment it provides wild plants and animals and the preservation of landraces. Through the provision of food, fodder, green manure, medicinal plants and material for cultural uses, the home gardens contribute to production services. The cultural list of services is extensive and includes spiritual, aesthetic and traditional aspects, a place for pastimes, environmental education and research, and a venue for perpetuation of heritage value, indigenous knowledge and social networks. There are other ecosystem services from home gardens. They are habitats for various microbial, insect, plant and animal species (Sthapit et al., 2006) and are refuges for biodiversity (Trinh et al., 2003). Seneviratne et al. (2010) discuss the nutrient cycling process in home gardens that continuously adds organic matter to soil. Prevention of soil erosion is also another benefit (Terra, 1954; Soemarwoto, 1987). The flowering plants in the garden attract pollinators, a service that is instrumental to crop production within and outside the home garden (Sthapit et al., 2006).

Limitations to home gardening

Literature on home gardens, while highlighting the numerous benefits, also recognizes a plethora of limitations to home gardening (Table 1.2). While households can cultivate a few crops in containers, those with land can do much more. As such, land availability and access are a major challenge to the development of home gardens. An associated issue is ownership or ownership-like rights that would encourage investment of time, effort and resources. Some commonly recognized constraints to home gardening are listed in Table 1.2.

Table 1.2 Limitations to home gardening adapted from Hoogerbrugge and Fresco (1993)

Resource and production issues	
Availability of quality inputs (e.g., seeds and livestock)	Brown et al. (1983), Engel et al. (1985), Ensing et al. (1985), Brownrigg (1985), Vasey (1985), Matahelumual and Verheul (1987), Best (1987, 1987), Singh (2019)
Access to land or lack of it and insecure land title	Thaman (1977), Engel et al. (1985), Ensing et al. (1985), Evers et al. (1985), Fernandes and Nair (1986), Solon (1988), Mitchell and Hanstad (2004)
Capital investments	Vasey (1985)
Time and/or labor	Engel et al. (1985), Best (1987), Solon (1988), Thaman (1977)
Crop losses	Okafor and Fernandes (1987)
Biotic stresses	
Pests and diseases outbreaks and roaming animals	Thaman (1977), Engel et al. (1985), Fernandes et al. (1984), Vasey (1985), Keatinge et al. (2012)
Abiotic stresses	
Soil quality, fertility and erosion	Thaman (1977), Bompard et al. (1980), Matahelumual and Verheul (1987), Keatinge et al. (2012)
Changes in climate and water availability	Niñez (1985a), Vasey (1985), Matahelumual and Verheul (1987), Buechler (2016)
Knowledge and skill gaps	
Knowledge, know-how and support from extension services	Thaman (1977), Ensing et al. (1985), Laumans et al. (1985), Best (1987), Matahelumual and Verheul (1987)
Research on home garden production	Evers et al. (1985)
Social and economic issues	
Theft	Thaman (1977), Engel et al. (1985), Fernandes et al. (1984), Vasey (1985), Keatinge et al. (2012)
Marketing issues	Ensing et al. (1985), Evers et al. (1985), Best (1987)
Cultural barriers and lack of information on nutritional benefits of crops	Miura (2003), Brun et al. (1989)

Source: Adapted from Hoogerbrugge and Fresco (1993).

Conclusions

Literature on home gardens has many perspectives to offer on the utility and versatility of home gardens and generally views them to be an eco-friendly and sustainable practice. Empirical evidence will continue to demonstrate various applications of home gardens, the benefits they provide, and the innovations that can be replicated or adapted. However, there is a dearth of scholarship related to the role of home gardens in conflict and post-conflict situations and analysis of the costs and benefits of home gardening. Expanding scholarship in those areas will strengthening the overall understanding of home gardens and demonstrate their applicability to diverse situations, for instance their use in disaster relief and management. More research is needed on economic aspects as well, especially with regards to quantifying net benefits and costs.

Despite its benefits and its reputation as low-input food production system, home gardening may not be as attractive or beneficial due to the various constraints listed earlier. Limitations to home gardening can discourage gardeners. As a result, they can become uninterested or completely withdraw from home gardening. Santhoshkumar and Ichikawa (2010) emphasize that low financial incentives can force home gardeners into activities that are economically beneficial but they may not always be environmentally sustainble.

This articulates the need to strengthen the policy areas focusing on home gardening as well as supporting institutions and organizations focusing on local food systems. Continued research can also play an important role in finding innovate ways to tackle these limitations and help maintain and sustain home gardens that can benefit and contribute toward the well-being of the household.

Notes

1 There are three forms of food insecurity: chronic food security is the most severe type, resulting from hunger or inability to consume the minimum amount of food needed for healthy life over a long period as a result of scarcity or poverty (FAO, 2008). Transitional (short-term) food insecurity is further subdivided into temporary (limited time period due to shocks) and seasonal or cyclical (trend) food insecurity.
2 Based on baseline projections, FAO (2006) reported an average consumption per person of 3,130 kcal per day by the year 2050. A lower average daily caloric availability per person of 3,047 kcal per day is estimated by Alexandratos (2009).

References

Abdoellah, O.S., Hadikusumah, H.Y., Takeuchi, K. and Okubo, S., 2006. Commercialization of homegardens in an Indonesian village: Vegetation composition and functional changes. *Agroforestry Systems*, 1(68), pp. 1–0013.
Abdoellah, O.S., Parikesit, Gunawan, B. and Hadikusumah, H.Y., 2001. *Home gardens in the Upper Citarum Watershed, West Java: A challenge for in situ conservation of plant genetic resources*. Witzenhausen, Germany: The International Plant Genetic Resources Institute, pp. 140–147.
Abebe, T., Wiersum, K., Bongers, F. and Sterck, F., 2006. Diversity and dynamics in homegardens of southern Ethiopia. In B.M. Kumar and P.K.R. Nair, eds. *Tropical homegardens: A time-tested example of sustainable agroforestry*. Dordrecht, The Netherlands: Springer Science, pp. 123–142.
Aceituno-Mata, L., 2010. *Estudio etnobotánico y agroecológico de la Sierra Norte de Madrid*. PhD Dissertation. Universidad Autónoma de Madrid, Madrid, Spain.
Agelet, A., Bonet, M.Á. and Vallès, J., 2000. Home gardens and their role as a main source of medicinal plants in mountain regions of Catalonia (Iberian Peninsula). *Economic Botany*, 54, pp. 295–309.

Aguilar-Støen, M., Moe, S.R. and Camargo-Ricalde, S.L., 2009. Home gardens sustain crop diversity and improve farm resilience in Candelaria Loxicha, Oaxaca, Mexico. *Human Ecology*, 37(1), pp. 55–77.

Albuquerque, U.P.A.L.H.C.C.J., 2005. Structure and floristics of homegardens in Northeastern Brazil. *Journal of Arid Environments*, 62, pp. 491–506.

Alexandratos, N., 2009. World food and agriculture to 2030/50: Highlights and views from mid-2009. In *How to feed the World in 2050. Proceedings of a technical meeting of experts*. Rome, Italy: Food and Agriculture Organization of the United Nations, pp. 1–32.

Ali, A.M.S., 2005. Home gardens in smallholder farming systems: Examples from Bangladesh. *Human Ecology*, 33, pp. 245–270.

Angel-Pérez, A.L. and Mendoza, M.A., 2004. Totonac homegardens and natural resources in Veracruz, Mexico. *Agriculture and Human Values*, 21, pp. 329–346.

Asfaw, Z., 2002. Home gardens in ethiopia: Some observations and generalizations. Proceedings of the Second International Home Gardens Workshop. Witzenhausen, Germany.

Bandarin, F., Hosagrahar, J. and Sailer Albernaz, F., 2011. Why development needs culture. *Journal of Cultural Heritage Management and Sustainable Development*, 1(1), pp. 15–25.

Bassullu, C. and Tolunay, A., 2010. General characteristics of traditional homegarden involving animal practices in rural areas of Isparta Region of Turkey. *Journal of Animal and Veterinary Advances*, 9, pp. 455–465.

Bennett-Lartey, S.O., Ayernor, G.S., Markwei, C.M., Asante, I.K., Abbiw, D.K., Boateng, S.K., Anchirinah, V.M. and Ekpe, P., 2002. Contribution of home gardens to in situ conservation of plant genetic resources farming systems in Ghana. In J.W. Watson and P.B. Eyzaguirre, eds. *Home gardens and in situ conservation of plant genetic resources in farming systems*. Italy, Rome: International Plant Genetic Resources Institute, p. 83.

Best, J., 1987. Homestead livestock and household livelihood in Sarawak: Innovations versus improvements. *Community Development Journal*, 3, pp. 197–201.

Bleasdale, T., Crouch, C. and Harlan, S.L., 2010–2011. Community gardening in disadvantaged neighborhoods in Phoenix, Arizona: Aligning programs with perceptions. *Journal of Agriculture, Food Systems, and Community Development*, 3(1), Winter, pp. 99–114.

Bompard, J., Ducatillion, C. and Heckersweiler, P.M.G., 1980. *A traditional agricultural system village forest gardens in West Java*. Montpellier, France: Academie de Montpellier.

Boserup, E., 1965. *The conditions of agricultural growth: The economics of Agrarian change under population pressure*. London: Allen & Unwin.

Brown, M. et al., 1983. Development of stabilized rainfed farming systems in the Intermediate Zone of Moneragala District, Sri Lanka. ICRA bulletin 14. International course for development oriented research in Agriculture. Wageningen, The Netherlands: Unpublished Mimeo.

Brownrigg, L., 1985. Home gardening. In *International development: What the literature shows*. Washington DC, USA: The League for International Food Education.

Brun, T., Reynaud, J. and Chevassus-Agnes, S., 1989. Food and nutritional impact of one home garden project in Senegal. *Ecology of Food and Nutrition*, 23(2), pp. 91–108.

Buchanan, R., 2012. *A Weaver's garden: Growing plants for natural dyes and fibers*. New York, USA ed. s.l.: Dover Publications, Inc.

Buchmann, C., 2009. Cuban home gardens and their role in social-ecological resilience. *Human Ecology*, 37, pp. 705–721.

Buechler, S., 2016. Gendered vulnerabilities and grassroots adaptation initiatives in home gardens and small orchards in Northwest Mexico. *Ambio*, 45(3), pp. 322–334.

Cai, C.T., Luo, J.S. and Nan, Y.Z., 2004. Energy and economic flow in homegardens in subtropical Yunnan, SW China: A case study on Sanjia village. *The International Journal of Sustainable Development & World Ecology*, 11(2), pp. 199–204.

Calvet-Mir, L., Gómez-Bagetthun, E. and Reyes-García, V., 2012. Beyond food production: Home gardens' ecosystem services: A case study in Vall Fosca, Catalan Pyrenees, northeastern Spain. *Ecological Economics*, 74, pp. 153–160.

Calvet-Mir, L. et al., 2016. Home garden ecosystem services valuation through a gender lens: A case study in the Catalan Pyrenees. *Sustainability*, 8(8), p. 718.

Christanty, L., Abdoellah, O., Marten, G. and Iskandar, J., 1986. Traditional agroforestry in West Java: the pekarangan (homegarden) and kebun-talun (annual-perennial rotation) cropping systems: Traditional agriculture in Southeast Asia. In G. Marten, ed. *Traditional agriculture in Southeast Asia*. Boulder, CO: Westview Press, pp. 132–158.

Cleveland, D., Orum, T.V. and Ferguson, N., 1985. Economic value of home vegetable gardens in an urban desert environment. *Hortscience*, 20(4), pp. 694–696.

Coomes, O.T. and Ban, N., 2004. Cultivated plant species diversity in home gardens of an Amazonian peasant village in northeastern Peru. *Economic Botany*, 58(3), pp. 420–434.

Danoesastro, H., 1980. *The role of homegardens as a source of additional daily income*. Paper presented at the Seminar on the Ecology of Homegardens III, Bandung, Indonesia.

Dash, S.S. and Misra, M.K., 2001. Studies on hill agro-ecosystems of three tribal villages on the Eastern Ghats of Orissa, India. *Agriculture, Ecosystems & Environment*, 86(3), pp. 287–302.

de la Cerda, H.E.C. and Mukul, R.R.G., 2008. Homegarden production and productivity in a Mayan community of Yucatan. *Human Ecology*, 36, pp. 423–433.

Del Angel-Pérez, A.L. and Alfonso, M.B.M., 2004. Totonac homegardens and natural resources in Veracruz, Mexico. *Agriculture and Human Values*, 21(4), pp. 329–346.

Drescher, A.W., Hagmann, J. and Chuma, E., 1999. Home gardens: A neglected potential for food security and sustainable land management in the communal lands of Zimbabwe. *Tropenlandwirt*, 100, pp. 163–180.

Engel, A. et al., 1985. Promoting smallholder cropping systems in Sierra Leone; an assessment of traditional cropping systems and recommendations for the Bo-Pujehun Rural Development Project. Seminar fuer Landwirtschaftliche Entwicklung (SLE). Centre for Advanced Training in Agricultural Development, Technische Universitaet Berlin. Fachbereich Internationalen Agrarentwicklung (FIA) Schriftenreihe des Fachbereichs nr. IV/86. Unpublished Mimeo.

Ensing, B., Freeks, G. and Sangers, S., 1985. *Homegardens and homegardening in the Matara district: The present situation and future prospects*. MSc thesis, Social Science and Economics Dept, University of Leiden, Netherlands.

Evers, G., Keleta, E., Kirway, T., Nikahetiya, S., Thom, L. and Vera, R., 1985. *A farming system study in the lowland wet zone of Sri Lanka, Agalawatta Division, Kalutara District*. Bulletin. Wageningen, Netherlands: International Centre for Development Oriented Research in Agriculture.

Eyzaguirre, P.B. and Linares, O.F., 2004. Introduction. In P.B. Eyzaguirre and O.F. Linares, eds. *Homegardens and agrobiodiversity*. Washington, DC, USA: Smithsonian Books, pp. 1–28.

Faber, M., Venter, S.L. and Benade, A.S., 2002. Increased vitamin A intake in children aged 2–5 years through targeted home-gardens in a rural South African community. *Public Health Nutrition*, 5(1), pp. 11–16.

Faber, M. and Wenhold, F., 2007. Nutrition in contemporary South Africa. *Water SA*, 33(3), pp. 393–400.

Ferdous, Z. et al., 2016. Development of home garden model for year round production and consumption for improving resource-poor household food security in Bangladesh. *NJAS: Wageningen Journal of Life Sciences*, 78, pp. 103–110.

Fernandes, E.C., Oktingati, A. and Maghembe, J., 1984. The Chagga homegardens: A multistoried agroforestry cropping system on Mt. Kilimanjaro (Northern Tanzania). *Agroforestry Systems*, 2(2), pp. 73–86.

Fernandes, E.C.M. and Nair, P.K.R., 1986. An evaluation of the structure and function of tropical homegardens. *Agricultural Systems*, 21, pp. 279–310.

Food and Agriculture Organization, 2006. *World agriculture: Towards 2030/2050.* Rome, Italy.

Food and Agriculture Organization, 2008. *An introduction to the basic concepts of food security.* Rome, Italy: EC-FAO Food Security Programme.

Food and Agriculture Organization, 2010. *The state of food insecurity in the world: Addressing food insecurity in protracted crises.* Rome, Italy.

Food and Agriculture Organization of the United Nations, 2009. *2050: A third more mouths to feed.* [Online] Available at: www.fao.org/news/story/en/item/35571/icode/ [Accessed 12 November 2019].

Fresco, L. and Westphal, E., 1988. A hierarchical classification of farm systems. *Experimental Agriculture*, 24(4), pp. 399–419.

Galhena, D.H., Mikunthan, G. and Maredia, K.M., 2012. Home gardens for enhancing food security in Sri Lanka. *Farming Matters*, Rio+20 Special Edition, p. 12.

Galluzzi, G., Eyzaguirre, P. and Negri, V., 2010. Home gardens: Neglected hotspots of agrobiodiversity and cultural diversity. *Biodiversity and Conservation*, 19(13), pp. 3635–3654.

Gautam, R. et al., 2009. Home gardens management of key species in Nepal: A way to maximize the use of useful diversity for the well-being of poor farmers. *Plant Genetic Resources*, 7(2), p. 14.

Gökçebağ, M. and Özden, Ö., 2017. Home garden herbs and medicinal plants of lefke, cyprus. *Indian Journal of Pharmaceutical Education and Research*, 51, pp. 441–444.

Gupta, A.K., 1989. Scientists' views of farmers' practices in India: Barriers to effective interaction. In R. Chambers, A. Pacey and L.A. Thrupp, eds. *Farmer first: Farmer innovation and agricultural research.* London, UK: Intermediate Technology Publications.

Helen Keller International-Asia-Pacific, 2010. Homestead food production model contributes to improved household food security, nutrition and female empowerment-experience from scaling-up programs in Asia (Bangladesh, Cambodia, Nepal and Philippines). *Nutrition Bulletin*, 8(1), March.

High, C. and Shackleton, C.M., 2000. The comparative value of wild and domestic plants in home gardens of a south African rural village. *Agroforestry Systems*, 48, pp. 141–156.

Holden, S.T., Hvoslef, H. and Simajuntak, R., 1996. Transmigration settlements in Seberida, Sumatra: Deterioration of farming system in a rain forest environment. *Agricultural Systems*, 49, pp. 237–258.

Hoogerbrugge, I. and Fresco, L.O., 1993. *Homegarden systems: Agricultural characteristics and challenges.* London, UK: International Institute for Environment and Development (IIED).

Huai, H. and Hamilton, A., 2009. Characteristics and functions of traditional homegardens: A review. *Frontiers of Biology in China*, 4(2), pp. 151–157.

Iannotti, L., Cunningham, K. and Ruel, M., 2009. *Improving diet quality and micronutrient nutrition: Homestead food production in Bangladesh.* Washington, DC, USA: International Food Policy Research Institute.

Ibnouf, F.O., 2009. The role of women in providing and improving household food security in Sudan: Implications for reducing hunger and malnutrition. *Journal of International Women's Studies*, 10(4), pp. 144–167.

Igwe, K., Agu-Aguiyi, F. and Nwazuruoke, G., 2014. Social and economic implications of home gardening on the livelihood of farm households in Abia State, Nigeria. *Developing Country Studies*, 4(1), pp. 66–71.

Janick, J., 2002. *Ancient Egyptian agriculture and the origins of horticulture.* Cairo, Egypt: ISHS Acta Horticulturae 582: International Symposium on Mediterranean Horticulture: Issues and Prospects, pp. 23–39.

Jesch, A., 2009. *Ethnobotanical survey of home gardens in Patones, Sierra Norte de Madrid, Spain: Management, use and conservation of crop diversity with a special focus on local varieties.* Thesis Dissertation, University of Natural Resources and Applied Life Sciences, Vienna.

Johnson-Welch, C. et al., 2000. *Improving household food security: Institutions, gender and integrated approaches.* s.l.: s.n.

Jose, D.N. and Shanmugaratnam, N., 1993. Traditional homegardens of Kerala: A sustainable human ecosystem. *Agroforestry Systems*, 24, pp. 203–213.

Karyono, I., 1990. Home gardens in Java: Structure and function. In K. Landauer and M. Brazil, eds. *Tropical home garden.* Tokyo: UN University Press, pp 138–147.

Keatinge, J.D., Chadha, M.L., Hughes, J.D.A., Easdown, W.J., Holmer, R.J., Tenkouano, A., Yang, R.Y., Mavlyanova, R., Neave, S., Afari-Sefa, V. and Luther, G., 2012. Vegetable gardens and their impact on the attainment of the Millennium Development Goals. *Biological Agriculture & Horticulture*, 28(2), pp. 71–85.

Kehlenbeck, K., Arifin, H.S. and Maass, B.L., 2007. Plant diversity in homegardens in a socio-economic and agro-ecological context. In T. Tscharntke et al., eds. *The stability of tropical rainforest margins, linking ecological, economic and social constraints of land use and conservation.* Berlin, Germany: Springer, pp. 297–319.

Kehlenbeck, K. and Maass, B.L., 2004. Crop diversity and classification of homegardens in Central Sulawesi, Indonesia. *Agroforestry Systems*, 63, pp. 53–62.

Kirabo, A., Byakagaba, P., Buyinza, M. and Namaalwa, J., 2011. Agroforestry as a land conflict management strategy in Western Uganda. *Environmental Research Journal*, 5(1), pp. 18–24.

Krishna, G.C., 2006. Home gardening *as a household nutrient garden.* Pokhara, Nepal; Local Initiatives for Biodiversity Research and Development, Bioversity International and Swiss Agency for Development and Cooperation, pp. 48–52.

Kumar, B.M. and Nair, P.K.R., 2004. The enigma of tropical homegardens. *Agroforestry Systems*, 61, pp. 35–152.

Landauer, K. and Brazil, M., 1990. *Tropical homegardens.* Tokyo, Japan: United Nations University Press.

Langellotto, G.A., 2014. What are the economic costs and benefits of home vegetable gardens? *Extension Journal: Research in Brief*, 52(2), pp. 1–7.

Laumans, Q., Kasijadi, F., Leksono, S., Nusantoro, B. and Santoso, P., 1985. *The home gardens of East Java: Results of an agro-economic survey.* Malang, Indonesia: Malang Research Institute for Food Crops.

Maroyi, A., 2009. Traditional home gardens and rural livelihoods in Nhema, Zimbabwe: A sustainable agroforestry system. *International Journal of Sustainable Development and World Ecology*, 16(1), pp. 1–8.

Marsh, R., 1998. Building on traditional gardening to improve household food security. *Food, Nutrition and Agriculture*, 22, pp. 4–14.

Matahelumual, M.M. and Verheul, M.A., 1987. *Vegetables in homegardens on East Java.* Wageningen: Scriptie Vakgroep Tropische PLantenteelt, LUW.

Mendez, V.E., Lok, R. and Somarriba, E., 2001. Interdisciplinary analysis of homegardens in Nicaragua: Micro-zonation, plant use and socioeconomic importance. *Agroforestry Systems*, 51, pp. 85–96.

Michon, G. and Mary, F., 1994. Conversion of traditional village gardens and new economic strategies of rural households in the area of Bogor, Indonesia. *Agroforestry Systems*, 25, pp. 31–58.

Midmore, D., Niñez, V. and Venkataraman, R., 1991. Household gardening projects in Asia: Past experience and future directions. *AVRDC Technical Bulletin*, 19.

Millat-e-Mustafa, M., Teklehaimanot, Z. and Haruni, A., 2002. Traditional uses of perennial homestead garden plants in Bangladesh. *Forests Trees Livelihoods*, 12, pp. 235–256.

Mitchell, R. and Hanstad, T., 2004. *Small homegarden plots and sustainable livelihoods for the poor.* Rome, Italy: Food and agriculture organization of the United Nations.

Miura, S., Osamu, K. and Susumu, W., 2003. Home gardening in urban poor communities of the Philippines. *International Journal of Food Sciences and Nutrition*, 54(1).

Moreno-Black, G., Somansang, P. and Thamathawan, S., 1996. Cultivating continuity and creating change: Woman's home garden practices in Northeastern Thailand. *Agriculture and Human Values*, 13(3), pp. 3–11.

Morton, L.W., Bitto, E.A., Oakland, M.J. and Sand, M., 2008. Accessing food resources: Rural and urban patterns of giving and getting food. *Agriculture and Human Values*, 25, pp. 107–119.

Nair, P.K.R., 1993. *An introduction to agroforestry*. Dordrecht, Netherlands: Kluwer Academic Publishers.

Nguyen, H. et al., 2017. *Understanding gender and power relations in home garden activities: Empowerment and sustainable home garden uptake*. Publication Number 17–813 ed. Taiwan: World Vegetable Center.

Niñez, V.K., 1985. *Household gardens: Theoretical considerations on an old survival strategy* (Vol. 1). Lima, Peru: International Potato Center.

Niñez, V.K., 1987. Household gardens: Theoretical and policy considerations. *Agricultural Systems*, 23(3), pp. 167–186.

Ochse, J., 1937. Horticulture and its importance in diet research. *Landbouw*, 13(in Dutch with English summary), pp. 202–225.

Ochse, J.J. and Terra, G.J.A., 1934. The function of money and products in relation to native diet and physical condition in Koetowinangoen (Java). 1. The agricultural and economic conditions of the natives and their food consumption. *Landbouw*, 10(4 and 5).

Okafor, J.C. and Fernandes, E.C., 1987. Compound farms of southeastern Nigeria. *Agroforestry Systems*, 5(2), pp. 153–168.

Okigbo, B., 1990. Home gardens in tropical Africa. In K. Landauer and M. Brazil, eds. *Tropical home gardens*. Tokyo, Japan: United Nations University Press, pp. 21–40.

Patalagsa, M., Schreinemachers, P., Begum, S. and Begum, S., 2015. Sowing seeds of empowerment: Effect of women's home garden training in Bangladesh. *Agriculture & Food Security*, 4(1), p. 24.

Perera, A.H. and Rajapakse, R.M.N., 1991. A baseline study of Kandyan forest gardens of Sri Lanka: Structure, composition and utilization. *Forest Ecology and Management*, 45, pp. 269–280.

Perrault-Archambault, M. and Coomes, O.T., 2008. Distribution of agrobiodiversity in home gardens along the Corrientes River, Peruvian Amazon. *Economic Botany*, 62(2), p. 109.

Peyre, A., Guidal, A., Wiersum, K.F. and Bongers, F., 2006. Dynamics of homegarden structure and functions in Kerala, India. *Agroforestry Systems*, 66, pp. 101–115.

Pulami, R. and Paudel, D., 2004. Contribution of home gardens to livelihoods of Nepalese farmers. In R. Gautam, B.R. Sthapit and P. Shrestha, eds. *Home Gardens in Nepal: Proceeding of a workshop on Enhancing the contribution of home garden to on: Farm management of plant genetic resources and to improve the livelihoods of Nepalese farmers: Lessons learned and policy implications*. Pokhara, Nepal: LI-BIRD, Bioversity International, and Swiss Agency for Development and Cooperation, pp. 18–26.

Pushpakumara, D. et al., 2012. A review of research on homegardens in Sri Lanka: The status, importance and future perspective. *Tropical Agriculturist*, 160, pp. 55–125.

Rammohan, A., Pritchard, B. and Dibley, M., 2019. Home gardens as a predictor of enhanced dietary diversity and food security in rural Myanmar. *BMC Public Health*, 19(1), p. 1145.

Ranasinghe, T.T., 2009. *Manual of low/no-space agriculture cum: Family business gardens*. AN Leusden, The Netherlands: RUAF Foundation.

Reyes-García, V., Vila, S., Aceituno-Mata, L., Calvet-Mir, L., Garnatje, T., Jesch, A., Lastra, J.J., Parada, M., Rigat, M., Vallès, J. and Pardo-de-Santayana, M., 2010. Gendered home gardens: A study in three mountain areas of the Iberian Peninsula. *Economic Botany*, 64, pp. 235–247.

Rigat, M., Garnatje, T. and Vallès, J., 2009. Estudio etnobotánico del alto valle del río Ter (Pirineo catalán): resultados preliminares sobre la biodiversidad de los huertos familiares [Ethnobotanical study of the high river Ter valley (Pyrenees): Preliminary results on bio-diversity in home gardens]. In F. Llamas and C. Acedo, eds. *Botánica PirenaicoCantábrica en el siglo XXI*. León, Spain: Universidad de León, pp. 399–408.

Rowe, W.C., 2009. "Kitchen gardens" in Tajikistan: The economic and cultural importance of small-scale private property in a post-soviet society. *Human Ecology*, 37, pp. 691–703.

Salam, M.A., Babu, K.S. and Mohanakumaran, N., 1995. Home garden agriculture in Kerala revisited. *Food and Nutrition Bulletin*, 16(3), pp. 1–4.

Santhoshkumar, A. and Ichikawa, K., 2010. Homegardens: Sustainable land use systems in Wayanad, Kerala, India. In C. Bélair, K. Ichikawa, B.Y.L. Wong and K.J. Mulongoy, eds. *Sustainable use of biological diversity in socio-ecological production landscapes: Background to the Satoyama Initiative for the benefit of biodiversity and human well-being*. Montreal, Canada: Secretariat of the Convention on Biological Diversity, pp. 125–128.

Schreinemachers, P. et al., 2015. The effect of women's home gardens on vegetable produc-tion and consumption in Bangladesh. *Food Security*, 7(1), pp. 97–107.

Seeth, H.T., Chachnov, S. and Surinov, A., 1998. Russian poverty: Muddling through eco-nomic transition with garden plots. *World Development*, 26(9), pp. 1611–1623.

Seneviratne, G., Kuruppuarachchi, K.A.J.M., Somaratne, S. and Seneviratne, K.A.C.N., 2010. Nutrient cycling and safety-net mechanism in the tropical Home gardens. *International Journal of Agricultural Research*, 5(7), pp. 529–542.

Singh, A., 2019. Traditional home gardens: Enhancing food security and livelihood in North East India. In R. Naresh, ed. *Research trends in agriculture sciences*. New Delhi, India: AkiNik Publications, pp. 59–72.

Soemarwoto, O., 1987. Homegardens: A traditional agroforestry system with a promising future. *Agroforestry: A decade of development*, pp. 157–170.

Soemarwoto, O. and Conway, G.R., 1991. The Javanese homegarden. *Journal for Farming Systems Research-Extension*, 2(3), pp. 95–118.

Solon, F.S., 1988. Food production through homegardening. *Gardening Nutritious Vegetables*. AVRDC Publication, No. 87–273.

Sthapit, B.R., Gautam, R. and Eyzaguirre, P., 2006. *The value of home gardens to small farmers*. Pokhara, Nepal: Local Initiatives for Biodiversity Research and Development, Bioversity International and Swiss Agency for Development and Cooperation, pp. 8–17.

Sthapit, B.R., Rana, R.B., Hue, N.N. and Rijal, D.R., 2004. The diversity of taro and sponge gourds in traditional home gardens in Nepal and Vietnam. In P.B. Eyzaguirre and O.F. Linares, eds. *Home gardens and agrobiodiversity*. Washington, DC, USA: Smithsonian Books, pp. 234–254.

Stoltzfus, R.J., Mullany, L. and Black, R.E., 2004. Iron deficiency anaemia. In E.M. Lopez, A.D., A. Rodgers, and C.J. Murray, eds. Comparative quantification of health risks. Global and regional burden of disease attributable to selected major risk factors. Geneva, Switzerland: World Health Organization, 2004, pp.1651–1801.

Sunwar, S., Thornström, C.G., Subedi, A. and Bystrom, M., 2006. Home gardens in western Nepal: Opportunities and challenges for on-farm management of agrobiodiversity. *Biodiversity & Conservation*, 15(13), pp. 4211–4238.

Talukder, A. et al., 2000. Increasing the production and consumption of vitamin A: Rich fruits and vegetables: Lessons learned in taking the Bangladesh homestead gardening programme to a national scale. *Food and Nutrition Bulletin*, 21(2), pp. 165–172.

Talukder, A. et al., 2006. *Homestead food production program in central and far: Western Nepal increases food and nutrition security: An overview of program achievements.* Pokhara, Nepal: Local Initiatives for Biodiversity Research and Development, Bioversity International and Swiss Agency for Development and Cooperation, pp. 27–34.

Tchatat, M., Puig, H. and Fabre, A., 1996. Genesis and organization of home gardens in the rainforest zone of Cameroun. *Revue D Ecologie – la Terre et la Vie*, 51, pp. 197–221.

Terra, G., 1954. Mixed-garden horticulture in Java. *Malaysian Journal of Tropical Geography*, 4, pp. 33–43.

Thaman, R.R., 1977. Urban root crop production in South West Pacific. Regional meeting on the production of root crops. Technical Paper – South Pacific Commission, 174, pp. 73–82.

Thaman, R.R., 1995. Urban food gardening in the Pacific Islands: A basis for food security in rapidly urbanising small-island states. *Habitat International*, 19(2), pp. 209–224.

Thompson, K., Austin, K.C., Smith, R.M., Warren, P.H., Angold, P.G. and Gaston, K.J., 2003. Urban domestic gardens (I): Putting small-scale plant diversity in context. *Journal of Vegetation Sciences*, 14, pp. 71–78.

Torquebiau, E., 1992. Are tropical agroforestry gardens sustainable? *Agriculture, Ecosystems and Environment*, 41, pp. 189–207.

Trinh, L.J.W.N.H. et al., 2003. Agrobiodiversity conservation and development in Vietnamese home gardens. *Agriculture, Ecosystems and Environment*, 97, pp. 317–344.

Vasey, D.E., 1985. Household gardens and their niche in Port Moresby, Papua New Guinea. *Food and Nutrition Bulletin*, 7(3), pp. 37–43.

Vazquez-Garcia, V., 2008. Gender, ethnicity, and economic status in plant management: Uncultivated edible plants among the Nahuas and Popolucas of Veracruz, Mexico. *Agriculture and Human Values*, 25(1), pp. 65–77.

von Baeyer, E., 2010. *The development and history of horticulture.* Oxford, United Kingdom: Eolss Publishers.

von Grebmer, K. et al., 2011. *The 2011 global hunger index.* Bonn, Germany, Washington, DC, USA, and Dublin, Ireland: International Food Policy Research Institute, Concern Worldwide and Welthungerhilfe.

Wanasundera, L., 2006. *Rural women in Sri Lanka's post-conflict rural economy.* s.l.: Centre for Women's Research and Food, Sri Lanka and Food and Agriculture Organization of the United Nations, Regional Office for Asia and the Pacific, Thailand.

Wezel, A. and Bender, S., 2003. Plant species diversity of homegardens of Cuba and its significance for household food supply. *Agroforestry Systems*, 57(1), pp. 39–49.

Wiersum, K.F., 2006. Diversity and change in homegarden cultivation in Indonesia. In B. Kumar and P. Nair, eds. *Tropical homegardens: A time-tested example of sustainable agroforestry.* Dordrecht, The Netherlands: Springer Science, pp. 13–24.

Wilkinson, A., 1994. Symbolism and design in ancient Egyptian gardens. *Garden History*, 22(1), pp. 1–17.

Wojtkowski, P.A., 1993. Toward an understanding of tropical home gardens. *Agroforestry Systems*, 24, pp. 215–222.

Yiridoe, E.K. and Anchirinah, V.M., 2005. Garden production systems and food security in Ghana: Characteristics of traditional knowledge and management systems. *Renewable Agriculture and Food Systems*, 20(3), pp. 168–180.

Zypchyn, K., 2012. Getting back to the garden: Reflections on gendered behaviours in home gardening. *Earth Common Journal*, 2(1), pp. 1–19.

2 Home gardens for nutritional security of men, women and children

Jacqueline d'Arros Hughes, John Donough Heber Keatinge and Julien Curaba

Introduction

About half of the world's population is malnourished due to over-consumption of specific food categories or under-consumption owing to a lack of food (World Health Organization (WHO), 2015a, 2015b). Almost two billion individuals are overweight or obese, yet nearly 800 million people, the equivalent of the population of Europe, are undernourished (FAO et al., 2012, 2015).

Children are the first noticeable victims of malnutrition. While overweight children are more likely to become obese adults and suffer from type 2 diabetes and other preventable non-communicable diseases, almost half of all children's deaths in 2011 (3.1 million children) were the consequence of undernutrition (Black et al., 2013).

Diets that do not contain the right balance of macronutrients and micronutrients are often associated with obesity or hunger brought about by excessive or insufficient consumption of carbohydrates, fat and proteins. However, unhealthy diets that lead to inadequate intake of essential minerals and vitamins are often overlooked as an important cause of non-communicable diseases. These diseases, such as anemia, blindness, cardiovascular disease, diabetes and cancer, kill around 38 million people each year, mostly in low- and middle-income countries (WHO, 2015c).

Dietary diversity and the role of fruit and vegetables in a balanced diet have been largely neglected in the drive to end the world's immediate hunger by ensuring food security through starch-based diets. Current agricultural research and development remains focused on carbohydrate-rich staples such as rice, wheat and maize that will not alone solve the difficulty many poor people face of eating a nutritious diet (as opposed to a high-calorie diet). There has not been enough investment directed toward the development of crops such as pulses, fruits and vegetables that are able to address mineral and vitamin deficiencies (Keatinge et al., 2011; Jamnadass et al., 2011). However, there has been a global shift by policy makers and donors recognizing the need to focus on nutrition as well as food security. In 2011, the Chicago Council on Global Affairs reported that agriculture needs to play a particularly important role in the prevention and mitigation of the growing "epidemic" of non-communicable diseases associated with poor and imbalanced diets (Nugent, 2011). This means that agriculture must focus on assuring the availability and affordability of fruit, vegetables, pulses and whole grains for balanced diets, which may contribute to a proportional reduction in the ingestion of excess carbohydrates, fats and salt.

The United Nations General Assembly declared 2014 the International Year of Family Farming and 2016 the International Year of Pulses, marking important steps toward the promotion of vegetable interventions in international agricultural research and development. Economic and agricultural growth should be "nutrition-sensitive," meaning that this growth needs to result in better nutritional outcomes through enhanced opportunities for the poor to diversify their diets (FAO et al., 2012). Agricultural programs such as home garden food production systems aim to promote locally available high-nutrient content vegetables and fruit, providing households with an affordable way to overcome malnutrition, especially micronutrient malnutrition, through diversified and balanced diets. These agricultural interventions, with the necessary market linkages, not only enhance diets but also can improve the incomes of smallholder family farmers – a key factor for reducing malnutrition.

A review of around 30 agricultural interventions that nominally influenced the nutritional status of consumers concluded that most interventions increased food production but did not necessarily improve nutrition or health within participating households (Berti et al., 2004). Home garden production was one of the positive exceptions noted. Bhutta et al. (2008) support the idea that dietary diversification intervention strategies, including the products of home gardening, are beneficial (Faber et al., 2002; Chadha and Oluoch, 2003), although evidence such as rigorous randomized control trials in support of such strategies is often lacking. Household or individual dietary diversity scores have been used as proxy indicators for dietary quality; attempts are increasing to develop trials where statistical evidence can be found on the effect of home gardening or fruit and vegetable consumption and the health of households or consumers.

Although scientific evidence for the effect of home gardens on nutrition has not yet been clearly established, scientists acknowledge the sustainable positive impact of home gardens on several outcomes including consumption, production, nutrition knowledge and women's empowerment. Home gardening has the potential to generate affordable and easy access to a healthy diversity of fruits and vegetables for householders. This type of agricultural intervention presents limited risk for households, is fairly easy to implement and can be a tool to raise awareness and tackle malnutrition simultaneously (Talukder et al., 2010).

An adaptable solution: easy to implement, low cost and ecologically sound

A home garden (also known as a nutrition garden or a homestead garden) is an area around the family dwelling where different vegetables, fruits, medicinal plants, herbs and spices are grown throughout the year, potentially with small ruminants, fowl or fish, to meet household nutritional requirements. They frequently are a system of production of diverse crop plant species and can be adjacent to or easily accessible to the household (Sunwar et al., 2006). Talukder et al. (2010) stated that the Helen Keller International "improved" model of home gardens should be maintained on fixed plots and should be able to provide a variety of fruits and vegetables throughout the year.

On a larger scale, gardens similar to home gardens can be used as school and community gardens. Such gardens often are encouraged where malnutrition has been identified as a serious problem for the disadvantaged, the young, the elderly or the sick. These gardens usually are of substantially greater size than most home gardens (De Neergard et al., 2009); they increase awareness of how food is produced and encourage an entrepreneurial spirit in a new generation of farmers.

Another type of home garden that has been proven beneficial is the disaster-recovery garden (AVRDC, 2011; Bhattarai et al., 2011; Luther and Lin, 2011; Yen, 2010). Developed by AVRDC to provide a fast and long-lasting solution for the victims of natural or human-made disasters, they are distributed in the form of vegetable seed packs. Each pack contains multiple hardy, fast-growing and nutritionally dense vegetable species that can be grown in small plots or containers. These species could include local, traditional vegetables which will vary according to location; for example amaranth (*Amaranthus* spp.), jute mallow (*Corchorus olitorious*) and African nightshade (*Solanum scabrum*) are some traditional vegetable species in Cameroon, which in combination will provide high levels of vitamins A, C and E as well as iron, folate and calcium (Yang and Keding, 2009). The varieties are chosen to be culturally acceptable, adapted to the local environment and open pollinated, thus allowing families to save seed for succeeding seasons. Instructions for successful production are included in appropriate local languages. These seed kits are usually distributed after the displaced families are settled in temporary or more permanent accommodation and have the time and energy to devote to vegetable gardening. If the plants are grown in containers, such as pots or polyethylene bags, these containers can be moved when the family has to move to a different location.

Home gardens are characterized by low capital input and simple production technologies. Land needs and input costs are small, and the potential use of household wastewater for irrigation, household organic wastes for compost and opportunities for vermiculture provide opportunities to make efficient use of limited resources, thus creating an ecologically sound system and reducing the overall cost of maintaining the garden. Home gardens are usually much more diverse than typical field-scale agriculture (Gari, 2003). Rotating crops in home gardens minimizes disease and pest inoculum, and the use of leguminous species in appropriate rotations enhances the soil health of the home garden.

Direct health benefits of home gardens

People who do not eat a balanced diet suffer from micronutrient deficiencies, among which iron, zinc and vitamin A deficiencies are the most common.

Micronutrient deficiencies, also known as hidden hunger, affect over 30% of the world's population, causing increased morbidity and mortality, impaired cognitive development and reduced learning ability and productivity (FAO et al., 2012). Deficiencies of vitamin A and zinc can result in death; deficiencies of iodine and iron cause stunting, which can contribute to children not achieving their developmental potential (Black et al., 2013).

Many papers record that the produce from home gardens can be of nutritional benefit to the families growing the gardens, especially for those family members that are malnourished and/or poor (Berti et al., 2004; Ehrenberg and Ault, 2005; Holmer, 2011; Keatinge et al., 2011; Niñez, 1985; Talukder et al., 2010). Home gardens can provide easy and regular access to a diversity of healthy vegetables and fruits, which may not be found in the local market, all year long.

A review by the UK Department for International Development (DFID) in 2014 presents a comprehensive and detailed analysis of nutrition outcomes from 14 home gardening studies. More than half of the studies show that home gardening is associated with increased household production and direct consumption of fruits and/or vegetables and an increase in the intake of foods rich in vitamin A (DFID, 2014; Girard et al., 2012). Food produced from home gardening can contain a substantial amount of micronutrients; in rural Bangladesh, home garden produce conferred a protective effect against night blindness among preschool children who had missed vitamin A supplementation (Campbell et al., 2011). Ten years after completion of a home garden project in a rural village in South Africa, it was clear that beta-carotene-rich vegetables and fruits continued to be planted and were more frequently consumed by the households that grew them (Zimpita et al., 2015).

Keatinge et al. (2011) demonstrate that in India, a 6 m² home garden can provide much of the vitamin requirements (vitamins A and C) for a family of four throughout the year. The introduction of gardens specifically designed by AVRDC to include nutrient-dense, well-adapted species – including pumpkin (*Cucurbita pepo*), tomato (*Solanum lycopersicum*), bittergourd (*Momordica charantia*) and cowpea (*Vigna unguiculata*) – for the benefit of all family members into a tribal community in Jharkhand State in India has proven to be of considerable nutritional value.

The inclusion of a range of indigenous or traditional vegetables brings further advantages, as many of these plants are nutrient-dense (Lin et al., 2009; Yang and Keding, 2009), often considerably more so than the so-called global vegetables currently available. They are often robust and easier to grow than some of the global vegetables. In the Pacific islands, such gardens usually include "slipperi kabis" or "bele" (*Abelmoschus manihot*), which has a very high level of folate, well outside the normal range (Keatinge et al., 2011). *A. manihot* samples at AVRDC have been shown to contain about 177 µg folate per 100 g, compared to some "global" tomatoes that have only 5 µg per 100 g of folate. Folate is a vital nutrient for pregnant women to assure proper development of the fetus.

An increase in home garden activities leads to an increase in the production and consumption of micronutrient-rich vegetables (Olney, 2009; Schreinemachers et al., 2015). Yet the evidence base for the impact of home gardening on the intake of micro- and macronutrients is still too small and inconsistent to draw definite conclusions (Ruel et al., 2013). Improving micronutrient intake through home gardening interventions should be a logical success, but patchy data on the relative bioavailability and bioconversion of nutrients from the vegetables that are consumed and a lack of replicated, statistically valid data means the specific impact of vegetables on the health of consumers is very difficult to document properly. Links between the agriculture and health sectors and protocols accepted by both sectors are needed to ensure the nutritional contribution from vegetables are properly characterized and quantified. Nevertheless, from health and medical information it is clear that dietary diversity (in the context of a healthy, balanced diet), is a strong proxy indicator to quantify the effect of home gardens and vegetable consumption on health.

In a joint report, the World Health Organization (WHO) and the United Nations Food and Agriculture Organization (FAO) stated that to achieve the best results in preventing non-communicable diseases, the strategies applied must fully recognize the essential roles of diet and nutrition as well as physical activity (WHO and FAO, 2003). Working in a home garden provides good physical exercise with a yield of home-grown produce as a reward. Yang et al. (2011) report that the total energy expended by a single female worker in a 6 m^2 garden over a three-month period (sowing to harvesting) was in the region of 54,000 kcal, and that this energy expenditure was more than that expended by regularly going to a gym as a form of exercise. In addition to being a potential source of healthy physical activity, home gardens also can be beneficial for the mind. In a 2014 study on the effect of horticultural therapy, participating in gardening around the home has been suggested as an effective treatment for mental and behavioral disorders (Kamioka et al., 2014).

The essential role of women: empowerment through home gardening

Home gardening is principally the responsibility of women throughout most of Africa, Latin America (Keys, 1999; Niñez, 1985; Pinton, 1985) and Asia (Afari-Sefa et al., 2012; Talukder et al., 2010), and it is a means of overcoming cultural and social restrictions on undertaking activities outside the home. Home gardens act as an empowering vehicle for women by providing opportunities to learn about ways to overcome malnutrition in their families and communities, the ability to grow the vegetables to help overcome malnutrition and, where there is excess production, to provide cash for additional household expenses (Mavlyanova, 2010; Talukder et al., 2000). Ruel and Levin (2001) contend that vegetable gardening enables women to have greater authority over the quality of the family diet.

Considering the importance of women in the selection of the family diet and their potential role in overcoming taboos and prejudices restricting the use of some nutritious vegetable species, empowering women through combined training in improved home gardens and nutrition makes an effective contribution for the success of home garden interventions. For example, social taboos play an important role in restricting vegetable consumption in Cameroon (Kamga et al., 2009). As Weller et al. (2011) indicated, although amaranth (*Amaranthus* spp.) is a very nutrient-dense and cheap vegetable in the Eldoret region of Kenya, its consumption is associated with poverty, and this may limit its marketability. In contrast, in western Kenya, this vegetable is commonly consumed, including by the wealthy, with other leafy greens such as African nightshade (e.g., *Solanum scabrum, S. villosum* and *S. nigrum*) to improve its taste. Such social pressures may be restricted culturally or geographically but nonetheless remain an important constraint to the inclusion of highly nutritious traditional vegetables in peri-urban or urban diets.

Nutritional impact on families and children is more likely when agricultural interventions target women and include women's empowerment activities (Ruel et al., 2013). Evaluation of many agricultural programs shows that women's health, social status, empowerment and control over resources are key mediators in the pathways between agriculture inputs, intra-household resource allocation and child nutrition (FAO, 2011; Ruel et al., 2013). Women tend to manage small vegetable plots; this requires time and effort, but it also means that with information and training the women are empowered to manage their household nutrition more efficiently and effectively. Assessments of home garden food production systems in Bangladesh and Nepal reported increases in women's income, in women's control over family resources and in their ability to influence in decision-making on a range of issues (Ruel et al., 2013).

In Bangladesh, 85% of women who participated in the Helen Keller International Homestead Gardens program stated that they had considerably increased their contribution to the household's resources and increased their status and decision-making power (Bushamuka et al., 2005). Similar findings were recorded by Bhattarai et al. (2011) in Indonesia, where women who had taken up vegetable production in Aceh Province claimed that they had new, extensive decision-making power in agricultural operations concerning vegetables. More recently, a significant positive impact was found through women's training in improved home gardens in poor rural households in Bangladesh (Schreinemachers et al., 2015). Strengthening women's control over the home garden led to increased supply and greater consumption of a diverse range of vegetables, thereby directly contributing to improved nutritional security for the whole family.

Gender issues are of critical importance in developing and adapting vegetable gardens, as species selection and prioritization are influenced by whether the principal gardener is male or female (HKI/Cambodia, 2003; Jamnadass et al., 2011). However, no studies have specifically accessed the consequences of home gardening on the empowerment of both women and men. In Kenya, a project promoting orange-fleshed sweet potato production among women farmers showed that while women gained control over selling the product, men maintained control over the income (Ruel et al., 2013).

Part of the benefits attributed to women's empowerment could be more widely associated with a general increase in the well-being of all family members. More than just a means to improve the family diet, home gardens are a place for mothers to educate their children about food production and the natural environment. Studies in Guatemala of the impact of Kaqchikel Maya gardens show that it is through work in the garden that mothers teach their children about plants and the care needed to tend fruit and vegetable crops and produce a good harvest (Keys, 1999).

Despite the benefits of women's empowerment through home gardening, very few studies have measured the effect of agriculture interventions on women's time, knowledge, practices, health or nutritional status and none have modeled the potential mediating role of these maternal resources on child nutrition (Ruel et al., 2013).

Community and school gardens, a promising approach to foster healthy behavior and an interest in horticulture

School gardens can be living laboratories and provide opportunities for children to explore the taste and texture options of vegetables. They are a potent means of empowering children to improve their health status. Evidence of such outcomes was provided by a project implemented by AVRDC in the Philippines (Asian Development Bank [ADB], 2006; Hoenicke et al., 2006). In Kenya, the government implemented school feeding programs in which school gardens, mainly planted with fruits and vegetables, have been a noted asset (Slow Food International, 2011).

In a recent assessment in North America of the effect of a school gardening program on Aboriginal First Nations children, the study showed a significant increase in the children's interest in vegetables and fruits over a seven-month period (Triador et al., 2015). School interventions have the potential to increase children's preferences for vegetables and fruits. The most promising strategies for improving generally low fruit and vegetable consumption levels among children are through local school food service policies (Ganann et al., 2014). Meals at school canteens can be supplemented with produce from school gardens; although the gardens usually cannot supply the total fruit and vegetable needs of the children, they can encourage children to develop an interest in vegetable and fruit gardening and to take the interest and seeds home to their families. Having a home garden was positively associated with children's diet diversity and with the frequency of vegetable consumption in Filipino preschools (Cabalda et al., 2011).

The Kitchen Garden National Program (SAKGNP), which has been providing garden and kitchen classes for children three to six years old in Australia, has proven to be a successful health promotion program with high community network impacts (Eckermann et al., 2014). This program spread to New Zealand, where school gardens have been acknowledged as a vehicle for extensive opportunities for health promotion (Dawson et al., 2014).

School and community gardens have important educational and advocacy functions not only for the young but also for all members of the community and may have further sociological benefits (Chen and Huang, 2005; Tenkouano, 2011). Vegetable gardens can provide opportunities for the young, the disabled and the elderly to participate in an active social life, have physical exercise and be productive (Adil, 1994; Chen and Huang, 2005; Keys, 1999; Koura et al., 2009). Such gardens have a special role in contributing to better integration of immigrant families into their new communities (Bellows et al., 2009). Communal garden projects in the southern Philippines that were linked with school feeding programs decreased malnutrition rates, provided additional income, strengthened overall community values, increased communal collaboration, contributed to enhanced self-esteem and further empowered the gardeners (Hill, 2011; Hobson and Hill, 2013).

The efficient use of water and organic wastes provides opportunities for advice to be given by health and sanitation professionals on ways to improve human health as well as the nutrient content of the garden soil (Holmer, 2011) by incorporating compost as well as properly treated human waste into the soil. The link between vegetable gardens and sustainable sanitation is a neglected research area. In many developing countries, more than half of the public elementary school population is infected with intestinal worms, a condition associated with malnutrition (Stephenson et al., 2000; Walker et al., 2007). Holmer and Monse (2006) described a Philippine school health program that successfully complemented basic hygiene interventions (daily handwashing and tooth brushing as well as biannual deworming) with a school garden component. The WASH (water, sanitation and hygiene) program, as described in the UNICEF WASH Strategy Paper (UNICEF, 2006), is a vital mechanism to ensure the health of children in relation to agriculture and school gardens. Contact with the soil and the consumption of agricultural and horticultural produce can be a risk if the school does not have access to and use of safe water and basic sanitation services and the mandate to promote improved hygiene among the children.

Home gardens: a local change leading to wider impact

Since home gardens are small in scale, do not pose a high risk to women and the household and offer direct benefits to householders, they are likely to proliferate as a "good way of living" within the community. Positive feedback from householders can quickly spread to neighbors. In South Africa, households still planting beta-carotene-rich vegetables and fruits ten years after a project completed were perceived as "well-to-do" and "healthy" households and as "givers" (Zimpita et al., 2015). The success of home gardens relies on their ability to be integrated into family routines and to create a stimulating nutritional environment. Promoting home consumption may require the participation of all the family, as well as the community, to raise awareness and increase the availability of produce for consumption.

In AVRDC's worldwide experience, most home and community gardeners will either sell or exchange some of their fresh produce with neighbors and friends (Mavlyanova, 2010). This can enhance the range of vegetables consumed by any specific family and can provide some vital supplemental income. The municipality of Opol (in northern Mindanao, Philippines) went from having one of the highest to one of the lowest malnutrition rates in the region (Hill, 2011) after ten years as a result of a community gardening and school feeding program in the municipality (D. Yasay, Mayor of Opol, personal communication, July 1, 2010), which engendered community well-being and a positive connection with the community garden through supply of surplus produce. In Jharkhand, India (Hariharan, 2010), some farmers who began with home gardens discovered new markets for leafy vegetables and decided to expand their production. AVRDC and its partners worked with several hundred young unemployed men and women in the Arumeru district in Tanzania to develop the necessary production and marketing skills to become small-scale commercial growers (Fintrac Inc., 2011). Household surveys and focus group discussions among 100 growers in an urban community garden project in the southern Philippines showed that about 70% of the vegetables produced were sold (Holmer and Drescher, 2006). In Cameroon, a survey of 300 peri-urban farmers in 2008 around the capital Yaoundé (Kamga et al., 2009) indicated that around 80% of the respondents both sold and consumed the produce of their vegetable plots. A study of home gardens in Bangladesh found that households earned an average of USD 8 bimonthly from selling fruits and vegetables, and the main use of this income was to buy extra food and to invest in other income-generating activities (Talukder et al., 2000). In a further example, home gardens in the Lima slums in Peru may provide earnings of up to USD 300 for an average family over a five-month period – an added indirect income of almost 10% from home garden produce (Niñez, 1985).

Proper treatment of vegetables is needed before they are consumed. Some vegetables are eaten raw, such as lettuce and tomato, and these must be carefully washed in clean water before consumption. It is critical that vegetables are properly cooked to ensure that the nutrients in the vegetables remain in the food after cooking and are bioavailable after consumption. Small changes in food preparation methods can make great improvements in bioavailability. For example, cooking mung bean with some tomato (raising the acidity) increases the bioavailability of iron or steaming leafy greens for a short time rather than extensive boiling will help retain the nutrients within the vegetables rather than being leached into the cooking water.

Addressing the problem of malnutrition through the development of school gardens is not only a solution for helping developing countries but also could be a successful solution to tackle obesity. Recently, Martinez and colleagues (2015) reported the successful implementation of a "sprouts" nutrition, cooking and gardening program for Latino youth on Los Angeles area school campuses. It is clear that the type of school garden must be adapted to each location and need, but whether the school garden is planted on spare land, restricted to containers, or is simply a small-scale intervention producing sprouts, each will contribute knowledge and develop interest in nutrition and good health.

Agricultural interventions including home and school gardening are not new strategies, as they have historically been present in most countries. However, they are relatively new strategies in the context of agricultural development with a focus on nutritional security rather than only food security. The effectiveness of the various interventions is well recognized, but a consensus needs to be built to reduce disparities and provide equitable access. It will be important to develop harmonized methods and analytical tools to be able to compare the data from different home and school gardens to be able to understand the dynamics and processes which lead to improved nutrition. This will certainly require further research to be able to impact the nutritional status of whole populations (Bhutta et al., 2013; Masset et al., 2012).

Discussion: how to improve the impact of home gardens?

Good agricultural practices, effective post-harvest management and hygienic and informed cooking methods, associated with education and training, are essential to ensure the sustainable success of home garden interventions. Future home gardening interventions will need to integrate appropriate varieties of traditional and global crops to combat local environmental constraints, new knowledge on raising high-quality seedlings, including grafting for soil-borne diseases and flood tolerance and information on correcting fertilization during establishment to boost productivity.

Breeding programs will have to continue to produce varieties of globally important and indigenous or traditional vegetables with enhanced biotic and abiotic stress resistance and improved nutritional and marketable traits. Although pest and disease resistant varieties often are not available for underutilized crops, simple agricultural practices such as the use of net shelters, crop rotation, water management and field sanitation can give sufficient protection to ensure a good harvest.

Careful selection of vegetable varieties with high nutrient content that fit into year-round cropping plans can address the need for a continuous supply of nutritious food. Garden plans should combine fast-growing crops for one-time harvest and annual and perennial crops that can be harvested multiple times. Development of recipes to ensure vegetable cooking methods enhance nutrient retention and bioavailability are important. West et al. (2002) call for new varieties of plants with enhanced vitamin A content and activity. Varieties of global vegetables currently produced for urban and peri-urban consumption and distributed through supermarkets have been bred for consistency in size, color, long shelf life and sometimes with new characteristics to tempt the purchaser (e.g., different/unexpected colors, smaller sizes of vegetables such as cauliflower that are usually large). Many of these traits come at the expense of nutritional quality. Davis et al. (2004) have shown that the nutritional content of many commercially grown vegetable varieties in North America has declined since the 1950s. Whether this is due to mass production techniques, the need to incorporate pest and disease resistance or enhance post-harvest qualities such as shelf life, it is apparent that the new varieties of these vegetables do contain a lower level of micronutrients and vitamins than their predecessors. There is a need to breed vegetables that are more nutritious, such as improved tomatoes with higher beta carotene content (Yang et al., 2007).

Good post-harvest knowledge is useful. The perishable nature of many vegetables, particularly in the tropics and especially green leafy types, means that they should be picked fresh when required, preserved through solar drying, pickling and bottling, or further processed through production of sauces, fermentation and so forth (Chadha and Oluoch, 2003). These mechanisms can be used to maintain and extend the shelf life of nutrient-dense vegetables prior to consumption (Habwe et al., 2008).

Producing healthy vegetables is of no benefit if they are not incorporated in the diet. Better practices in the kitchen will help to ensure that the nutritional value of vegetables is maximized for consumers (AVRDC, 2010). Development of recipes incorporating vegetables for tasty and attractive meals is an extremely important and neglected area. In many countries, the school curriculum no longer includes traditional home economics, which would give the students a good grounding in many of the issues discussed in this paper: production of food around the house, how to store it, to prepare it properly and to ensure it is attractive and tasty to consume. Ngegba et al. (2008) have shown that iron availability was increased significantly when sweet potato leaves were prepared using improved recipes compared with traditional preparations. Similarly, Vijayalakshmi et al. (2003) found that the available iron from mung bean (*Vigna radiata*) for adolescent schoolgirls can be substantially improved when the beans are prepared with green leafy vegetables, tomatoes and some vegetable oil. This resulted in measurable improvements in hemoglobin levels, body mass index and health in the study population. Special recipes that enhance mung bean iron availability have been developed for use in northern Indian cuisine (Bains et al., 2003).

The potential use of household wastewater for irrigation and household organic wastes for compost provides opportunities to make efficient use of limited resources. This is of benefit to the environment and can become a sustainable household production system (BMVBS/BBR, 2008; Torquebiau, 1992). However, the specific guidelines for the use of wastewater, as defined by the WHO and by national governments, must be translated into local protocols that best suit the agronomic requirements of the crops grown as well as the specific socioeconomic, cultural and environmental realities of each targeted country (WHO, 2007; Seidu et al., 2008).

Home gardens have an unappreciated role in the vegetable marketing chain. One of the key constraints to vegetable research and development activities is the lack of real production data. Some information is available and collected and compiled by the FAO to give an indication of the production of, for example, tomatoes. However, this data does not take into account the large amount of tomatoes that are produced on a very small scale, either for home consumption or for sale in local wet markets. Thus, the FAO data is probably a considerable underestimate. Additionally, although tomatoes and some other crops are singled out, the remaining crops are consolidated under "vegetables," a term that encompasses many thousands of vegetables consumed around the world. A further complication for the researcher is the inherent difficulty of data collection from home gardens. Most women who tend home gardens will not be in a position to quantify their daily harvest of a few leaves here, one fruit there – although the impact on the family's diet may be substantial. Although researchers tend to agree that there is a lack of evidence on the effect of home gardens on improving nutritional status, they concur that this may be due to research design issues rather than the absence of a true effect (DFID, 2014; Girard et al., 2012). Fully integrated research, including rigorous impact assessment, is needed to understand the role of home gardens and gardeners in income generation, post-harvest handling and value addition, crop hygiene, product contamination issues and marketing strategies.

Positive changes already have taken place in several countries. For example, governments in the Philippines and Uzbekistan have adopted a series of decrees to promote the establishment of school gardens in public schools as part of national educational campaigns to reduce hunger and malnutrition (Philippine Presidential Executive Order 776–2009 and Administrative Order 5–2011; Government of Uzbekistan Decree–2010; Bondarchuk, 2011). These policy changes will have considerable development impact that will need to be assessed over the next decade.

More integrated efforts will provide the necessary evidence from the agriculture, nutrition, health and education sectors to demonstrate the effectiveness and sustainability of this critical pro-poor development area. With accurate and pertinent data, the task of national advocacy for a supportive policy environment will become much easier. Only then will the potential impact of home gardening really be seen and properly understood.

References

Adil, J.R., 1994. *Accessible gardening for people with physical disabilities: A guide to methods, tools and plants.* Bethesda, MD, USA: Woodbine House.

Afari-Sefa, V., Tenkouano, A., Ojiewo, C.O., Keatinge, J.D.H. and Hughes, J.D.A., 2012. Vegetable breeding in Africa: Constraints, complexity and contributions toward achieving food and nutritional security. *Food Security*, 4(1), pp. 115–127.

Asian Development Bank, 2006. *Final project report for RETA 6067.* Manila, Philippines: Asian Development Bank.

Bains, K., Yang, R.Y. and Shanmugasundaram, S., 2003. *High-iron mungbean recipes for North India* (Vol. 3, No. 562). Manila, Philippines: AVRDC and World Vegetable Center.

Bellows, A.C., Alcaraz, G.V. and Vivar, T., 2009. Gardening as tool to foster health and cultural identity in the context of international migration: Attitudes and constraints in a female population. *II International Conference on Landscape and Urban Horticulture*, 881, June, pp. 785–792.

Berti, P.R., Krasevec, J. and FitzGerald, S., 2004. A review of the effectiveness of agriculture interventions in improving nutrition outcomes. *Public Health Nutrition*, 7(5), pp. 599–609.

Bhattarai, M., Fitriana, N., Ferizal, M., Luther, G.C., Mariyono, J. and Wu, M.H., 2011. *Vegetables for improving livelihoods in disaster-affected areas: A socioeconomic analysis of Aceh, Indonesia*. Shanhua, Taiwan: AVRDC and The World Vegetable Center.

Bhutta, Z.A., Ahmed, T., Black, R.E., Cousens, S., Dewey, K., Giugliani, E., Haider, B.A., Kirkwood, B., Morris, S.S., Sachdev, H.P.S. and Shekar, M., 2008. What works? Interventions for maternal and child undernutrition and survival. *The Lancet*, 371(9610), pp. 417–440.

Bhutta, Z.A., Salam, R.A. and Das, J.K., 2013. Meeting the challenges of micronutrient malnutrition in the developing world. *British Medical Bulletin*, 106(1), pp. 7–17.

Black, R.E., Victora, C.G., Walker, S.P., Bhutta, Z.A., Christian, P., De Onis, M., Ezzati, M., Grantham-McGregor, S., Katz, J., Martorell, R. and Uauy, R., 2013. Maternal and child undernutrition and overweight in low-income and middle-income countries. *The Lancet*, 382(9890), pp. 427–451.

Bondarchuk, R., 2011. School nourishment: Food and health. *Uzbekistan National News Agency*, [internet] November. Available at: www.uza.uz/ru/society/17105 [Accessed 22 November 2011].

Bundesministerium für Verkehr, Bau und Stadtentwicklung/Bundesamt für Bauwesen und Raumordnung, 2008. Städtebauliche, ökologische und soziale Bedeutung des Kleingartenwesens. In *Forschungen Heft 133*. Bonn, Germany: Bundesministerium für Verkehr, Bau und Stadtentwicklung/Bundesamt für Bauwesen und Raumordnung.

Bushamuka, V.N., de Pee, S., Talukder, A., Kiess, L., Panagides, D., Taher, A. and Bloem, M., 2005. Impact of a homestead gardening program on household food security and empowerment of women in Bangladesh. *Food and Nutrition Bulletin*, 26(1), pp. 17–25.

Cabalda, A.B., Rayco-Solon, P., Solon, J.A.A. and Solon, F.S., 2011. Home gardening is associated with Filipino preschool children's dietary diversity. *Journal of the American Dietetic Association*, 111(5), pp. 711–715.

Campbell, A.A., Akhter, N., Sun, K., De Pee, S., Kraemer, K., Moench-Pfanner, R., Rah, J.H., Badham, J., Bloem, M.W. and Semba, R.D., 2011. Relationship of homestead food production with night blindness among children below 5 years of age in Bangladesh. *Public Health Nutrition*, 14(9), pp. 1627–1631.

Chadha, M.L. and Oluoch, M.O., 2003. Home-based vegetable gardens and other strategies to overcome micronutrient malnutrition in developing countries. *Food Nutrition and Agriculture*, (32), pp. 17–23.

Chen, H. and Huang, Y., 2005. The theory and application of horticultural therapy. *Journal of the Chinese Society for Horticultural Science*, 51(2), pp. 135–144.

Davis, D.R., Epp, M.D. and Riordan, H.D., 2004. Changes in USDA food composition data for 43 garden crops, 1950 to 1999. *Journal of the American College of Nutrition*, 23(6), pp. 669–682.

Dawson, A., Richards, R., Collins, C., Reeder, A.I. and Gray, A., 2014. Edible gardens in early childhood education settings in Aotearoa, New Zealand. *Health Promotion Journal of Australia*, 24(3), pp. 214–218.

De Neergaard, A., Drescher, A.W. and Kouamé, C., 2009. Urban and peri-urban agriculture in African cities. In C.M. Shackleton, M.W. Pasquini and A.W. Drescher, eds. *African indigenous vegetables in urban agriculture*. London, UK: Earthscan, pp. 67–96.

Department For International Development, 2014. Can agriculture interventions promote nutrition? In *Agriculture and nutrition evidence paper*. London, UK: DFID.

Eckermann, S., Dawber, J., Yeatman, H., Quinsey, K. and Morris, D., 2014. Evaluating return on investment in a school based health promotion and prevention program: The investment multiplier for the Stephanie Alexander Kitchen Garden National Program. *Social Science & Medicine*, 114, pp. 103–112.

Ehrenberg, J.P. and Ault, S.K., 2005. Neglected diseases of neglected populations: Thinking to reshape the determinants of health in Latin America and the Caribbean. *BMC Public Health*, 5(1), p. 119.

Faber, M., Phungula, M.A., Venter, S.L., Dhansay, M.A. and Benadé, A.S., 2002. Home gardens focusing on the production of yellow and dark-green leafy vegetables increase the serum retinol concentrations of 2–5-y-old children in South Africa. *The American Journal of Clinical Nutrition*, 76(5), pp. 1048–1054.

Fintrac Inc., 2011. *Tanzania agricultural productivity program*. [Online] (Updated 2015) Available at: www.fintrac.com/projects/tanzania#collapseOne [Accessed 10 April 2005].

Food and Agriculture Organization of the United Nations, 2011. *The state of food insecurity in the world 2011: How does international price volatility affect domestic economies and food security?* Rome, Italy: Food and Agriculture Organization of the United Nations.

Food and Agriculture Organization of the United Nations, World Food Program, and International Fund for Agricultural Development, 2012. *The state of food insecurity in the world 2012: Economic growth is necessary but not sufficient to accelerate reduction of hunger and malnutrition*. Rome, Italy: Food and Agriculture Organization of the United Nations.

Food and Agriculture Organization of the United Nations, World Food Program, and International Fund for Agricultural Development, 2015. *The state of food insecurity in the world 2015: Meeting the 2015 international hunger targets: Taking stock of uneven progress*. Rome, Italy: Food and Agriculture Organization of the United Nations.

Ganann, R., Fitzpatrick-Lewis, D., Ciliska, D., Peirson, L.J., Warren, R.L., Fieldhouse, P., Delgado-Noguera, M.F., Tort, S., Hams, S.P., Martinez-Zapata, M.J. and Wolfenden, L., 2014. Enhancing nutritional environments through access to fruit and vegetables in schools and homes among children and youth: A systematic review. *BMC Research Notes*, 7(1), p. 422.

Gari, J.A., 2003. Agrobiodiversity strategies to combat food insecurity and HIV. In *AIDs impact in rural Africa*. Rome, Italy: Food and Agriculture Organization of the United Nations.

Girard, A.W., Self, J.L., McAuliffe, C. and Olude, O., 2012. The effects of household food production strategies on the health and nutrition outcomes of women and young children: A systematic review. *Paediatric and Perinatal Epidemiology*, 26, pp. 205–222.

Habwe, F.O., Walingo, K.M. and Onyango, M.O.A., 2008. Food processing and preparation technologies for sustainable utilization of African indigenous vegetables for nutrition security and wealth creation in Kenya. In G.L. Robertson and J.R. Lupien, eds. *Using food science and technology to improve nutrition and promote national development*. Oakville, Ontario: IUFoST.

Hariharan, S., 2010. Business planning to discover the future scope of AVRDC intervention to larger partners of CINI. Mumbai (India): SRTT, Project report. Home garden NBJK. *SRTT Foundation Internal Report*. Mumbai, India: SRTT.

Helen Keller International/Cambodia, 2003. *Handbook for home gardening in Cambodia: The complete manual for vegetable and fruit production*. Phnom Penh, Cambodia: Helen Keller International.

Hill, A., 2011. A helping hand and many green thumbs: Local government, citizens and the growth of a community-based food economy. *Local Environment*, 16(6), pp. 539–553.

Hobson, K. and Hill, A., 2013. Cultivating citizen-subjects through collective praxis: Organized gardening projects in Australia and the Philippines. In T. Lewis and E. Potter, eds. *Ethical consumption: A critical introduction.* Abingdon, UK: Taylor and Francis, pp. 216–230.

Hoenicke, M., Ecker, O., Qaim, M. and Weinberger, K., 2006. Iron and vitamin A consumption and the role of indigenous vegetables: A household level analysis in the Philippines. *Discussion Paper No. 03/2006.* Institute of Agricultural Economics and Social Sciences in the Tropics and Subtropics. Stuttgart, Germany: University of Hohenheim.

Holmer, R.J., 2011. Vegetable gardens benefit the urban poor in the Philippines. *Appropriate Technology*, 38(2), pp. 49–51.

Holmer, R.J. and Drescher, A.D., 2006. Promouvoir la sécurité alimentaire: Le rôle des jardins familiaux [Improving food and nutritional security: The role of family gardens]. *Magazine Agriculture Urbaine*, 15, pp. 17–18.

Holmer, R.J. and Monse, B., 2006. School health programmes pay dividends. *Appropriate Technology*, 33(4), pp. 56–59.

Jamnadass, R.H., Dawson, I.K., Franzel, S., Leakey, R.R.B., Mithöfer, D., Akinnifesi, F.K. and Tchoundjeu, Z., 2011. Improving livelihoods and nutrition in sub: Saharan Africa through the promotion of indigenous and exotic fruit production in smallholders' agroforestry systems: A review. *International Forestry Review*, 13(3), pp. 338–354.

Kamga, R., Kouame, C. and Akyeampong, E., 2009. Vegetable consumption patterns in Yaoundé, Cameroon. *Annals of Nutrition Metabolism*, 55(supplement 1), p. 613.

Kamioka, H., Tsutani, K., Yamada, M., Park, H., Okuizumi, H., Honda, T., Okada, S., Park, S.J., Kitayuguchi, J., Abe, T. and Handa, S., 2014. Effectiveness of horticultural therapy: A systematic review of randomized controlled trials. *Complementary Therapies in Medicine*, 22(5), pp. 930–943.

Keatinge, J.D.H., Yang, R.Y., Hughes, J.D.A., Easdown, W.J. and Holmer, R., 2011. The importance of vegetables in ensuring both food and nutritional security in attainment of the Millennium Development Goals. *Food Security*, 3(4), pp. 491–501.

Keys, E., 1999. Kaqchikel gardens: Women, children, and multiple roles of gardens among the Maya of highland Guatemala. *Yearbook, Conference of Latin Americanist Geographers*, 25, January, pp. 89–100.

Koura, S., Oshikawa, T., Ogawa, N., Snyder, S.M., Nagatomo, M. and Nishikawa, C., 2009. Utilization of horticultural therapy for elderly persons in the urban environment. *ISHS Acta Horticulturae 881: II International Conference on Landscape and Urban Horticulture*, June, pp. 865–868.

Lin, L.-J., Hsiao, Y.-Y. and Kuo, C.- G., 2009. Promising indigenous vegetables. In *Discovering indigenous treasures: Promising indigenous vegetables from around the world*. Shanhua, Taiwan: World Vegetable Center.

Luther, G.C. and Lin, L.J., 2011. Rural appraisal of the impacts from the Typhoon Morakot seed distribution. *AVRDC Feed Field*, 9, pp. 5–6.

Martinez, L.C., Gatto, N.M., Spruijt-Metz, D. and Davis, J.N., 2015. Design and methodology of the LA Sprouts nutrition, cooking and gardening program for Latino youth: A randomized controlled intervention. *Contemporary Clinical Trials*, 42, pp. 219–227.

Masset, E., Haddad, L., Cornelius, A. and Isaza-Castro, J., 2012. Effectiveness of agricultural interventions that aim to improve nutritional status of children: Systematic review. *British Medical Journal*, 344, p. d8222.

Mavlyanova, R., 2010. Vegetable systems in central Asia and the Caucasus: Research and development to improve livelihood security. In *Horticulture and livelihood security*. Jodhpur, India: Scientific Publishers, pp. 2026–2034.

Ngegba, J.B., Msuya, J.M. and Yang, R.Y., 2008. In vitro iron bioavailability in sweet potato leaf recipes as affected by processing methods. *International Symposium on Under-utilized Plants for Food Security, Nutrition, Income and Sustainable Development*, 806, March, pp. 385–390.

Niñez, V., 1985. Working at half-potential: Constructive analysis of home garden programmes in the Lima slums with suggestions for an alternative approach. *Food and Nutrition Bulletin*, 7(3), pp. 1–9.

Nugent, R., 2011. *Bringing agriculture to the table: How agriculture and food can play a role in preventing chronic disease*. Chicago, IL, USA: The Chicago Council on Global Affairs.

Olney, D.K., Talukder, A., Iannotti, L.L., Ruel, M.T. and Quinn, V., 2009. Assessing impact and impact pathways of a homestead food production program on household and child nutrition in Cambodia. *Food and Nutrition Bulletin*, 30(4), pp. 355–369.

Pinton, F., 1985. The tropical garden as a sustainable food system: A comparison of Indians and settlers in Northern Colombia. *Food and Nutrition Bulletin*, 7(3), pp. 1–4.

Ruel, M.T., Alderman, H. and Maternal and Child Nutrition Study Group, 2013. Nutrition-sensitive interventions and programmes: How can they help to accelerate progress in improving maternal and child nutrition? *The Lancet*, 382(9891), pp. 536–551.

Ruel, M.T. and Levin, C.E., 2001. Discussion paper 92: Assessing the potential for food-based strategies to reduce vitamin A and iron deficiencies: A review of recent evidence. *Food and Nutrition Bulletin*, 22(1), pp. 94–95.

Schreinemachers, P., Patalagsa, M.A., Islam, M.R., Uddin, M.N., Ahmad, S., Biswas, S.C., Ahmed, M.T., Yang, R.Y., Hanson, P., Begum, S. and Takagi, C., 2015. The effect of women's home gardens on vegetable production and consumption in Bangladesh. *Food Security*, 7(1), pp. 97–107.

Seidu, R., Heistad, A., Amoah, P., Drechsel, P., Jenssen, P.D. and Stenström, T.A., 2008. Quantification of the health risk associated with wastewater reuse in Accra, Ghana: A contribution toward local guidelines. *Journal of Water and Health*, 6(4), pp. 461–471.

Slow Food International, 2011. *School gardens in Kenya food education project*. [Online] Available at: http://confolio.vm.grnet.gr/scam/23/resource/4.

Stephenson, L.S., Latham, M.C. and Ottesen, E.A., 2000. Malnutrition and parasitic helminth infections. *Parasitology*, 121(S1), pp. S23–S38.

Sunwar, S., Thornström, C.G., Subedi, A. and Bystrom, M., 2006. Home gardens in western Nepal: Opportunities and challenges for on-farm management of agrobiodiversity. *Biodiversity & Conservation*, 15(13), pp. 4211–4238.

Talukder, A., Haselow, N.J., Osei, A.K., Villate, E., Reario, D., Kroeun, H., SokHoing, L., Uddin, A., Dhunge, S. and Quinn, V., 2010. Homestead food production model contributes to improved household food security and nutrition status of young children and women in poor populations. Lessons learned from scaling-up programs in Asia (Bangladesh, Cambodia, Nepal and Philippines). *Field Actions Science Reports: The Journal of Field Actions*, Special Issue 1.

Talukder, A., Kiess, L., Huq, N., De Pee, S., Darnton-Hill, I. and Bloem, M.W., 2000. Increasing the production and consumption of vitamin A: Rich fruits and vegetables: Lessons learned in taking the Bangladesh homestead gardening programme to a national scale. *Food and Nutrition Bulletin*, 21(2), pp. 165–172.

Tenkouano, A., 2011. The nutritional and economic potential of vegetables. In L. Starke, ed. *State of the world 2011: Innovations that Nourish the planet*. Norton and New York, USA: Worldwatch Institute.

Torquebiau, E., 1992. Are tropical agroforestry home gardens sustainable? *Agriculture, Ecosystems & Environment*, 41(2), pp. 189–207.

Triador, L., Farmer, A., Maximova, K., Willows, N. and Kootenay, J., 2015. A school gardening and healthy snack program increased Aboriginal First Nations children's preferences toward vegetables and fruit. *Journal of Nutrition Education and Behavior*, 47(2), pp. 176–180.

United Nations Children's Fund, 2006. *UNICEF water, sanitation and hygiene annual report.* Rome, Italy: UNICEF.

Vijayalakshmi, P., Amirthaveni, S., Devadas, R.P., Weinberger, K., Tsou, S.C.S. and Shanmugasundaram, S., 2003. *Enhanced bioavailability of iron from mungbeans and its effects on health of schoolchildren.* Shanhua, Taiwan: AVRDC and World Vegetable Center.

Walker, S.P., Wachs, T.D., Gardner, J.M., Lozoff, B., Wasserman, G.A., Pollitt, E., Carter, J.A. and International Child Development Steering Group, 2007. Child development: Risk factors for adverse outcomes in developing countries. *The Lancet*, 369(9556), pp. 145–157.

Weller, S.C., Marshall, M.L., Pitchay, D., Ngouaijo, M., Obura, P., Ndinya, C. and Ojiewo, C., 2011. Development of sustainable African indigenous vegetable production and market-chain for small-holder farmers in Kenya and Tanzania. *Horticultural Collaborative Research Support Program Project Report*. Davis, CA, USA: University of California, Davis.

West, C.E., Eilander, A. and van Lieshout, M., 2002. Consequences of revised estimates of carotenoid bioefficacy for dietary control of vitamin A deficiency in developing countries. *The Journal of Nutrition*, 132(9), pp. 2920S–2926S.

World Health Organization, 2007. *Guidelines for the safe use of wastewater, excreta and greywater* (Vol. 1). Geneva, Switzerland: World Health Organization.

World Health Organization, 2015a. *Obesity and overweight.* [Online] (Updated January 2005) Available at: www.who.int/mediacentre/factsheets/fs311/en/.

World Health Organization, 2015b. *Micronutrient deficiencies.* [Online] (Updated January 2005) Available at: www.who.int/nutrition/topics/ida/en/.

World Health Organization, 2015c. *Noncommunicable diseases.* [Online] (Updated January 2005) Available at: www.who.int/mediacentre/factsheets/fs355/en.

World Health Organization and Food and Agriculture Organization of the United Nations, 2003. Diet, nutrition and the prevention of chronic diseases. *WHO Technical Report Series, No. 916*. Geneva, Switzerland: World Health Organization.

Yang, R.-Y., Hanson, P.M. and Lumpkin, T.A., 2007. Better health through horticulture: AVRDC's approach to improved nutrition of the poor. *Acta Horticulturae*, 744, pp. 71–77.

Yang, R.-Y., Huang, Y.-C. and Shiao, Y.-Y., 2011. Physical, nutritional and environmental benefits of working in an urban vegetable garden. *Asia Pacific Conference on Clinical Nutrition*. Bangkok, Thailand, 5–9 June.

Yang, R.-Y. and Keding, G.B., 2009. Nutritional contributions of important: African indigenous vegetables. In C.M. Shackleton, M.W. Pasquini and A.W. Drescher, eds. *African indigenous vegetables in urban agriculture*. London, UK: Earthscan, pp. 105–143.

Yen, M.H., 2010. 240 days after the devastating earthquake: Agricultural rehabilitation in Central Haiti. *AVRDC Feed Field*, 7, pp. 3–4.

Zimpita, T., Biggs, C. and Faber, M., 2015. Gardening practices in a rural village in South Africa 10 years after completion of a home garden project. *Food and Nutrition Bulletin*, 36(1), pp. 33–42.

3 Keeping it close to home

Home gardens and biodiversity conservation

Gamini Pushpakumara, Jessica Sokolow, Bhuwon Sthapit, Wawan Sujarwo and Danny Hunter

Introduction

Home gardens represent a dense, dynamic and multistoried arrangement of mixed but compatible species harboring high levels of biodiversity. People grow a wide variety of vegetables, fruits, root and tuber crops, fodder and medicinal plants as well as other economically important plant species providing spices, flowers and building materials. Livestock, poultry, small fishponds and bees can also be found in many home gardens. In addition to this high inter-specific diversity, home gardens are important repositories for significant intra-specific biodiversity, especially of plant genetic resources, and they are often home to important and unique traditional varieties and landraces (Soemarwoto, 1987; Pushpakumara et al., 2012, 2016). In fact, home gardens are often sites of many species, rare and unique varieties and landraces that are not commonly found in larger fields or the wider farming system (Shrestha et al., 2004). Further, farmer experimentation is common in home gardens, and the composition and structure of plant and animal species found is largely a result of farmer evaluation and selection, exchange and management (Eyzaguirre and Linares, 2004). Home gardens, therefore, represent an important element of any strategy to conserve biodiversity.

This chapter will discuss the determinants of biodiversity in these microclimates, including agro-ecological, socio-economic, cultural and political factors which can create better enabling environments for the role of home gardens. While these factors have the potential to foster biodiversity, they also can threaten its existence. One principal threat relates to market pressures, which can drive home garden owners to commercialize their practices, shifting production from traditional plants to commercial species, some of which are more easily cultivated and marketable and adopting new management practices that can erode genetic resources and drive biodiversity loss (Vlkova et al., 2011; Kahane et al., 2013; Caballero-Serrano et al., 2016). These kinds of hazards to conservation efforts have not gone unnoticed. However, positive policy support to home gardens for family well-being and women's empowerment linked to unique agricultural biodiversity can counteract these hazards.

The global community has drawn attention to biodiversity loss and genetic erosion, putting the need to conserve the world's biodiversity firmly on national and international multi-sector agendas. Home gardens as a tool for biodiversity conservation falls well within these established international and national strategies and plans. Though others argue that the size of home gardens are so small that they cannot be considered a viable unit of any conservation strategy, we argue that it is indeed possible for home gardens to play this role especially by creating networks of home gardens spread across landscapes and communities, thus serving as meta-populations, combined with strengthening informal social seed networks and local seed markets. The Convention on Biological Diversity (CBD) adopted a *Strategic Plan for Biodiversity 2011–2020 and the Aichi Targets* in 2010 that established an international framework and associated objectives for biodiversity. This international strategy calls for the engagement of national policy makers through National Biodiversity Strategies and Actions Plans (NBSAPs) to address issues of biodiversity loss and genetic erosion. Even before this 2010 strategic plan, the CBD put in place the Global Strategy for Plant Conservation (GSPC) in 2002 to prevent the further loss of plant diversity. Along with these strategies and plans, the goals for the 2030 Agenda for Sustainable Development incorporate the need to protect and promote biodiversity on both the land (Target 15) and sea (Target 14), and Target 2.5 is specific to maintaining genetic diversity (see Box 3.1 for specific descriptions of the targets) as part of an overall goal promoting sustainable agriculture, food security, improving nutrition and ending hunger. These agendas, strategies and plans set forth international and national targets that create space for home gardens in meeting biodiversity conservation objectives but also the achievement of multiple other benefits highlighted elsewhere in this book (see Box 3.1 for highlighted targets). However, home gardens as sites and spaces for biodiversity conservation require further attention, particularly in this context of the 2030 Agenda for Sustainable Development.

Box 3.1 Relevant international targets to home gardens and biodiversity conservation

CBD's Aichi Targets

Target 7: "By 2020 areas under agriculture, aquaculture and forestry are managed sustainably, ensuring conservation of biodiversity."

Target 13: "By 2020, the genetic diversity of cultivated plants and farmed and domesticated animals and of wild relatives, including other socio-economically as well as culturally valuable species, is maintained and strategies have been developed and implemented for minimizing genetic erosion and safeguarding their genetic diversity" (United Nations Environment Programme/CBD, 2010).

CBD's GSPC

Target 6: "At least 75 percent of production lands in each sector managed sustainably, consistent with the conservation of plant diversity."

Target 9: Seventy percent of the genetic diversity of crops including their wild relatives and other socio-economically valuable plant species conserved, while respecting, preserving and maintaining associated indigenous and local knowledge (CBD, 2012).

Sustainable Development Goals

Goal 2, Target 2.5: "By 2020 maintain genetic diversity of seeds, cultivated plants, farmed and domesticated animals and their related wild species, including through soundly managed and diversified seed and plant banks at national, regional and international levels and ensure access to and fair and equitable sharing of benefits arising from the utilization of genetic resources and associated traditional knowledge as internationally agreed" (United Nations, undated).

Home gardens serve as unique and particularly effective entry points for biodiversity conservation and use. They often promote both in situ and ex situ conservation, which prevents the loss of biodiversity through its maintenance within natural and human-made ecosystems, respectively. Home gardeners often do not cultivate agro-biodiversity in their backyards with the intention of serving as conservationists, though some recognize the importance of the wealth of ecosystem services they provide (Caballero-Serrano et al., 2016). While most home gardeners understand the need to protect their surrounding environment, the primary function of these spaces is to support their more immediate needs and values in their daily lives (Reyes-García et al., 2010; Pushpakumara et al., 2010, 2012; Kortright and Wakefield, 2011). Home gardens benefit households from all socioeconomic brackets, however in many corners of the world they prove to be important assets for poor and marginalized groups in society (Birol, 2004; Suwal et al., 2008; Galluzzi et al., 2010). Home gardens play an important role in contributing to food security, nutrition and dietary diversity (Gautam et al., 2009; Buchmann, 2009; Pushpakumara et al., 2010, 2012), as well as other aspects of well-being including income generation (Trinh et al., 2003; Salako et al., 2014; Poot-Pool et al., 2015) and medicinal, aesthetic, cultural and religious values (Trinh et al., 2003; Gautam et al., 2009; Buchmann, 2009; Pushpakumara et al., 2010, 2012; Kortright and Wakefield, 2011). These aspects of home gardens are largely dealt with in other chapters of this volume and will not be the focus of this chapter. As home gardeners cultivate and maintain their gardens with these goals in mind, biodiversity is promoted often as the element which binds all these multiple benefits.

This chapter will review the current literature on the role and importance of home gardens in the conservation and sustainable use of biodiversity using examples from most geographical regions of the world including the Global North and South, as well as rural and urban contexts. The chapter will also highlight two case studies from Indonesia and Sri Lanka and will bring to light principles for scaling-up the potential of home gardens for the benefit of on-farm biodiversity conservation.

Common features of home gardens

As home gardens are shaped by factors both within and outside the household, they often vary widely in content and form, with structure, composition and size determined largely by agro-ecology and local food culture; it is in part this variability that contributes to their importance in conservation. Yet through studies of home gardens throughout the world, they are also found to have many common features that help to define home gardens and understand their biodiversity contributions.

Physical and social characteristics of home garden plots

Home gardens are located within homesteads, allowing home gardeners to consistently tend to the plot and easily take advantage of its products (Galluzzi et al., 2010). The objectives of domesticating plant and animal species close to the homestead are multiple, including easy access, availability of fresh and quality products for the kitchen, availability of fresh produce over longer periods (perennial species) and family preferences. Home gardens are often distinguished from their surroundings by the use of fences, which contributes to the security of the household while also protecting the genetic resources. Composition and choice of species for this purpose are often tall fuel-wood trees, climbers and fodder or bio-fertilizing species with direct use value as well as for the protection of home garden produce. Many regions – from Europe to South East Asia – use "live fences" that delineate the property while still allowing for gene flow between the garden and the natural environment (Karyono, 1990). These fences can be made of hedges, tall trees, or other species that ensure crops within the plot have access to sufficient light resources (Watson and Eyzaguirre, 2002; Shrestha et al., 2004; Buchmann, 2009; Galluzzi et al., 2010).

Individually, each home garden tends to be small in size; the worldwide average size of home garden units is around 0.1–0.5 ha or 1,000–5,000 m^2 (Trinh et al., 2003). This small size is in part a consequence of outside pressures. For example, in an urban environment there is competition with buildings and infrastructure development, while in a rural environment the garden is impacted by the financial resources and land rights available to a household and are thus often proportional to the size of the overall farm (Guarino and Hoogendijk, 2004). In addition to the garden plot, these rural gardens may also have a separate location for post-harvesting activities, including preparing crops for use in the home or for sale in the market (Gautam et al., 2009). While the land areas of these production systems are small, these production systems collectively contribute important assets to environmental and human health that cannot be found in larger agricultural practices.

As one would expect, due to spatial constraints as well as sociocultural and economic factors impacting the landowners (discussed in greater depth later), there is a positive correlation between land size and species richness (Sunwar et al., 2006; Gautam et al., 2009). In many regions, expanding land size in an effort to further conservation efforts is not a realistic achievement. Other factors must be taken into consideration, such as the financial resources available to the farmers, which have been correlated to species richness in China (Yongneng et al., 2006), or land ownership rights, which can influence a home gardener's willingness to make long-term investments in the garden plot (Adhikari et al., 2004; Galluzzi et al., 2010; Das and Das, 2015). While financial and political constraints like these may limit the biodiversity present in garden plots, social networks may allow genetic diversity to flourish, thus shaping the physical constructs of the home garden (Shrestha et al., 2004; Buchmann, 2009; Galluzzi et al., 2010).

Many home gardens are carefully composed puzzles, with species intricately woven together giving way to high species diversity (Hemp, 2006; Norfolk et al., 2014). The carefully designed architecture – over space and time – makes way for diverse, curated environments. For example, if you were to walk through home gardens of Nepal, you would see trees both supporting the needs of the family as a sources of fuel and fodder, but also supporting other crops, as trees can serve as support to climbing crops (Shrestha et al., 2004). In some regions, trees have low intra-specific diversity due to their higher demand for space (Galluzzi et al., 2010). Spatial constraints are particularly seen in peri-urban or urban areas, where tree diversity is often low (Bernholt et al., 2009; Poot-Pool et al., 2015). Furthermore, surveys in Central Italy found that 40% of traditional fruit tree varieties are often present as individuals within single home gardens (Pavia et al., 2009; Galluzzi et al., 2010). In commercial areas in India over the last 30 years, home garden and mango orchards have been turning into multi-varietal orchards to avoid risk arising from climate change and market adversity (Gajanana et al., 2016). Thus to understand intra-specific diversity of these fruit tree species, a collection of home gardens must be considered (Galluzzi et al., 2010). These surveys raise awareness on the need for complementary in situ and ex situ strategies, which will be discussed in detail within this chapter.

The pieces of the puzzle are curated by the home gardeners who manage and determine the diversity of species in their garden based on their household's needs and preferences (discussed under "Individual Factors"). Home gardeners develop configurations that maximize the available space and further the presence of various species. They also provide unique habitats based on the idiosyncratic needs of the species and varieties, thus furthering species diversity by creating suitable niches and microclimates (Galluzzi et al., 2010; Pushpaku-mara et al., 2012; Cruz-Garcia and Struik, 2015).

Many home gardeners take advantage of the vertical as well as the horizontal space available in their plots. Different vertical arrangements can be found in gardens around the world and often depend on the resources available, as well as the knowledge and experience of the home gardener. For example, Balinese home gardens are characterized by their vertical structure and a large percent-age of fruit trees, which can also serve as a support for climbing plants (Sujarwo and Caneva, 2015; see Box 3.2 for a case study on Balinese home gardens). These arrangements take advantage of differentiated root structures, utilizing the nutrients from various soil levels, as well as taking advantage of different light requirements (Eyzaguirre and Linares, 2004), therefore supporting sus-tainable and resilient ecosystems (Smith et al., 2006; Weerahewa et al., 2012). In part, these complex arrangements allow home gardeners to maximize the number of species incorporated into their garden plans.

Box 3.2 Balinese home garden composition, species richness and plant use

Balinese home gardens represent a significant reservoir of genetic resources especially in remote villages, as well as annual and perennial plant species distributed in three-dimensional and temporal space. Balinese home gar-dens may also have a diversity of useful animals associated with them such as chickens, ducks, goats, cows and pigs. This maximizes production while meeting social and economic needs in the form of cash income, food security and medicines. Recent ethnobotanical studies of Balinese home gardens have demonstrated around 36 cultivated plant species belonging to 20 families and 29 genera, the most common families being Zingib-eraceae, followed by Poaceae, Fabaceae, Anacardiaceae, Cucurbitaceae, Asteraceae and Euphorbiaceae. Most plants are collected throughout the year and most frequently used parts are leaves (including young leaves), fruits, tuberous roots and young shoots. In a few cases a single plant part has multiple uses, e.g., the fruit juice of the uncooked ripe fruit of tamarillo (*Cyphomandra betacea* (Cav.) Miers.) used by locals for aphthous stomatitis and hypertension.

The home gardens in some aga (indigenous Balinese) villages are located in close vicinity to inhabited houses and composed mainly of fruit plants. The gardens provide plant materials for quick and easy access to foodstuffs, like white mango (*Mangifera caesia* Jack), fragrant mango (*Mangifera odorata* Griff.), purple mangosteen (*Garcinia mangostana* L.), avocado (*Persea americana* Mill.), pomegranate (*Punica granatum* L.) and Chinese mulberry (*Morus australis* Poir). These plants are produced in limited numbers for household and local consumption. Among the trees, the most valuable are durian (*Durio zibethinus* L.) for its fruit and timber, coconut (*Cocos nucifera* L) for its fruit and timber, *Toona sureni* (Blume) Merr. for its timber and its role as a shade tree, cinnamon (*Cinnamomum burmanni* (Ness & T. Ness) Blume), robusta coffee (*Coffea canephora* Pierre ex A. Froehner) and clove (*Syzygium aromaticum* (L.) Merr. & L.M. Perry), the last of which is particularly favored in the drier areas. The top five cultivated plants grown in Balinese home gardens are cassava (*Manihot esculenta* Crantz.), followed by banana (*Musa* × *parasidiaca* L.), sambong (*Blumea balsamifera* (L.) DC.), sweet potato (*Ipomoea batatas* (L.) Lam.) and papaya (*Carica papaya* L.). These species, with the exception of *Blumia balsamifera*, which is used to treat diarrhea, fever, heartburn and constipation, are the main food species consumed among the local people. The leaf of *Manihot esculenta*, which is the most commonly used of all plant species and is known by all age groups of the community, is particularly appreciated because it is easy to grow and can be used in a wide variety of dishes. One of Bali's most famous traditional foods is *jukut ares* soup, made from the core of the stems of the banana plant (*Musa* × *parasidiaca*).

Among the cultivated plants and categorized as local food crops, the most important market-oriented crops are salak/snake fruit (*Salacca zalacca* (Gaertn.) Voss), *Carica papaya*, ridged gourd (*Luffa acutangula* (L.) Roxb.), butter bean (*Phaseolus lunatus* L.), common bean (*Phaseolus vulgaris* L.), winged bean (*Psophocarpus tetragonolobus* (L.) DC) and rambutan (*Nephelium lappaceum* L.). *Ipomoea batatas* and *Manihot esculenta* are both used as food crops and are very valuable cash crops for the local people. In addition to food uses, some species are used as medicines to treat constipation, cough, diarrhea, dysentery, fever, headache, heartburn, hypertension, rheumatism, aphthous stomatitis, to relax the nerves, as a diuretic and to raise body heat. For example, the decoction of leaves of pandan (*Pandanus amaryllifolius* Roxb.) is used for rheumatism and to relax the nerves, while the juice of the tuberous roots of java ginger (*Curcuma zanthorrhiza* Roxb.), known in Balinese as *temu agung*, are used among the indigenous population for heartburn. Furthermore, the sap of sugar cane (*Saccharum officinarum* L.) is used locally for cough. The vertical structure of home gardens in Bali is due to the presence of a large number of trees, especially of the fruit species. Of the total number of cultivated plants, the traditional villages with the highest diversity contained 15 species, whereas those with the lowest had nine species. The high abundance of plant species found in the village is a result of the high number of individuals of some cultivated plants, i.e., aromatic ginger (*Kaempferia galanga* L.), peacock ginger (*Kaempferia rotunda* L.), *Morus australis*, turmeric (*Curcuma domestica* Valeton), *Ipomoea batatas*, *Manihot esculenta*, *Musa* × *paradisiaca*, ginger (*Zingiber officinale* Roscoe) and *Pandanus amaryllifolius*.

If conserved sustainably, these plants can be a good source of income for rural communities and strengthen food security and household health.

However, some use changes were observed and identified as causes of decline for yam (*Amorphophallus paeoniifolius* (Dennst.) Nicolson), sugar palm (*Arenga pinnata* (Wurmb) Merr.), palmyra palm (*Borassus flabellifer L.*), wild cherry (*Antidesma bunius* (L.) Spreng.), bilimbi (*Averrhoa bilimbi* L.) and many more. In extreme cases the gardens are given over to a single species – cloves, citrus or cassava – and all the traditional features of diversity, complexity, multiple use and stratification are lost.

Sources: Sujarwo et al. (2014), Sujarwo and Caneva (2015), Sujarwo et al. (2016)

Home for multiple species and varieties

Home gardeners may incorporate inter-specific and intra-specific diversity to contribute to family's needs. In subsistence households, this diversity can contribute to the family's food security and dietary diversity among other things. For those plants and animals not used in the household, families may bring them to the market to sell for cash or barter for other goods (Engels, 2002; Kehlenbeck and Maass, 2004; Shrestha et al., 2004; Sujarwo and Caneva, 2015). The diversity of species and varieties depends on a suite of interacting agro-ecological, socioeconomic, cultural, individual and political factors that – when considered together – have a unique impact on the intra- and inter-specific diversity found in each home garden. For example, Gautam et al. (2009) found that many home gardens of Nepal did not have much intra-specific diversity of fruit and fodder species, as they were often not essential to meeting a household's needs, whereas Kandyan home gardens (KHGs) in Sri Lanka are often sites of considerable intra-specific diversity (see Box 3.3 on KHGs in Sri Lanka). Additionally, in many urban cities within the Global North, intra-specific diversity is impacted by animal husbandry restrictions, often enacted under zoning, animal welfare and public health laws, which aim to keep livestock and other farm animals off of public property, control noise and smell, and provide adequate living conditions (Kortright and Wakefield, 2011; MRSC, n.d.). In Vietnam, the rearing of animals is an important feature of the home gardens, contributing to food security and improved financial capabilities of many rural families (Trinh et al., 2003).

Subsistence farmers in rural regions will manipulate inter- and intra-specific diversity to reduce risks and satisfy environmental conditions (Williams, 2004). As a result, home gardens serve as resource reservoirs in times of stress, including lean periods (Cruz-Garcia and Struik, 2015). Home gardeners will intentionally curate their plots with species of different life cycles and domestication status; for example, incorporating annual and perennial species (Gautam et al., 2009) as well as different post-harvest processing and storage capabilities ensures diverse sources of food will be available at different times throughout the year (Shrestha et al., 2004; Galluzzi et al., 2010).

Box 3.3 Kandyan home gardens and biodiversity conservation in Sri Lanka

Kandyan home gardens (KHGs) have been an integral part of the Sri Lanka's landscape and culture for centuries, and the composition and structure of plant and animal species found are the product of a combination of farmers' selection, natural evolution and environmental suitability. It is estimated that around 70% of the households in Kandy and adjacent districts have a long-standing home garden. These biocultural diverse sites blend characteristics which meet the socioeconomic, cultural and ecological needs of the region's diverse communities and landscapes. KHGs represent a dynamic land use system that over time and space maintains and even enhances and creates crop genetic diversity, which provides for a wide range of products year-round that helps increase household self-reliance. KHGs also contribute valuable ecosystem services and reduce pressure on fragmented natural forests by connecting them with a biodiversity-friendly land use system. KHGs contribute the provision of goods (food, spices, medicines, timber), regulating services (carbon sequestration, prevention of soil erosion), cultural services (an aesthetic environment, employment, social prestige) and supporting services (habitat for wild flora and fauna including pollinators, nutrient cycling).

It is estimated that about 50% of the species diversity of fruit crops in Sri Lanka is conserved in KHGs, with the most common fruit tree species being jackfruit, mango, cashew, citrus, guava, sweet orange, rambutan and avocado. One KHG is estimated to conserve 10 to 20 fruit crop species, and one or two individual trees per species may be sufficient to provide adequate fruit for household consumption. While individual populations of species in KHGs may be small, at a landscape level they are a vital refuge for plant and animal species that are neither grown or raised in the wider agro-ecosystems nor found in the wild. Field surveys with farmers have revealed that the majority of known fruit crop varieties in Sri Lanka are found in KHGs, highlighting their important value in the conservation of genetic diversity. Although few studies have been carried out on the genetic diversity of fruit crops in KHGs, morphological and genetic diversity assessments of jackfruit have revealed that much of the genetic variation of the species is conserved in KHGs. Further, observations of the jackfruit population in the Kandy district for fruiting season, fruit shape, number of fruits per tree, fruit weight, flesh thickness and hardness, flesh texture, aroma, colour and juiciness, and latex quantity has revealed wide variation. Similarly, distribution of mango morphotypes such as *gira amba* and *mee amba* suggests that the bulk of the genetic diversity of many perennial fruit crop species is conserved through KHGs. Clearly, KHGs in Sri Lanka constitute a valuable system for on-farm conservation of genetic diversity and facilitation of their gene flow. Further, home gardens, especially KHGs, also act as a repository of conservation of genetic resources of many spice crops, such as cloves, nutmeg, pepper, cardamom, allspice, cinnamon, betel leaf and betel nut.

However, recent observations reveal that certain varieties of fruit crops are being lost from the KHGs due to the promotion of improved varieties of grafted planting material, lack of attention and care, fragmentation due to population pressure and damage by animals such as monkeys, wild boars and porcupines. Despite this, there still remain a large number of KHGs at the landscape scale, which clearly helps safeguard local varieties and landraces of many fruit species. The importance of the KHG network in Sri Lanka as an element in a complementary conservation strategy for conservation of fruit crop genetic resources is critically important for Sri Lanka. KHGs also help conserve other crops, trees and livestock and poultry species as well as fruit crops in this unique landscape and contribute to wider biodiversity conservation and ecosystem services. The value of KHGs is clearly recognized and appreciated by households and communities. Overall, the number and total area of KHGs have been rising annually, despite little policy support. The value of this traditionally developed agroforestry system for addressing present and future challenges is increasingly recognized at a political level also. The national development policy framework of the government of Sri Lanka now includes strategies to expand and improve food production in such landscapes of the country, while the National Agriculture Policy of 2007 also highlights the need to promote their value.

Source: Pushpakumara et al. (2010, 2012, 2016).

Many of the plant species have multiple uses, demonstrating the importance of plants in home gardens for subsistence and as part of local culture and heritage (Williams, 2004; Sujarwo and Caneva, 2015). Folk naming systems recognize the multiple uses and serve as a tool to help understand farmers' preferences and perceptions, determining what they value most in the species (color, size, shape, taste, smell and so on) (Gessler and Hodel, 2004). Such characteristics – and consequentially their names – help to understand how biodiversity is managed in the home gardens (Gessler and Hodel, 2004). One species may have a diversity of names due to different perceptions. When curating the species in plots, some home gardeners may use methods that enable greater inter-specific diversity. For example, it is common practice in many home gardens in Vietnam to combine vegetables and fruits with fishponds and livestock (e.g., pigs and buffalo). Ponds, while making up a small percentage of landholdings, are an important component of home gardens in Vietnam. They also serve as an important protein supply and source of income for families. The ponds are used to recycle vegetable waste from home gardens and household refuse, thus feeding the fish in the ponds (Trinh et al., 2003).

Value of home garden biodiversity

Households rely on the diversity of home gardens to support their well-being in a variety of capacities, with crops often serving multiple purposes. Additionally, these values, whether food security, income generation or recreation, can act as an impetus for the selection of different species, varieties and landraces. Values may vary from household to household, which further contributes to the conservation potential of these unique, diverse habitats. When developing policies and programs, it important to take these variations into account (Gautam et al., 2009; Taylor and Lovell, 2014).

Contribution to food security, nutrition and dietary diversity

Home gardens serve as an important source of intra- and inter-specific biodiversity important to the food security, nutrition and dietary diversity of families, particularly marginalized families who may in part rely on the garden for subsistence (see Boxes 3.2 and 3.3 discussing home gardens in Bali and Sri Lanka, respectively; Trinh et al., 2003; Kehlenbeck and Maass, 2004; Cruz-Garcia and Struik, 2015). A study found that the presence of a home garden can have significant positive impacts on a child's nutritional status, with the biodiversity – not the size – being the most important factor (Jones et al., 2005). The food diversity found in these biodiverse production systems can often provide the breadth of nutrients and bioactive non-nutrients humans require (Hunter et al., 2015). These gardens rich in biodiversity augment food diversity available to families throughout the year by maximizing the species available (Trinh et al., 2003; Cruz-Garcia and Struik, 2015). Additionally, families often prefer the food grown in their home gardens due to its safety, freshness, taste and nutrition in comparison to food purchased from a market (Calvet-Mir et al., 2012; Freedman, 2015). Trinh and colleagues (2003) reported that that gardening households on average are supplied with more than 50% of their supply of vegetables, fruits, plantains and herbs from the gardens rich in diversity.

A whole diet approach can help to inform how these biodiverse microenvironments can contribute to improved health outcomes. This approach considers the diverse combinations of different foods and their interactions, as well as the cultural and/or religious traditions of a household, to meet an individual's nutritional needs (Hunter et al., 2015). Home gardens maintain a rich level of biodiversity found within and between species that contribute to an individual's whole diet, benefiting their dietary diversity and nutritional outcomes. Some households obtain additional nutritional benefits from incorporating new species – such as fish – into their diets. In Bangladesh, where inland water bodies and ecosystems maintain diverse species and varieties of fish, many households are now expanding freshwater aquaculture. These biodiverse aquaculture production systems incorporated in home garden production systems around the world provide essential and bioavailable nutrients, including important, high-quality protein and fatty acids (Hunter et al., 2015).

The intra-specific diversity found in home gardens also has the potential to yield significant nutritional benefits. Scientific research shows that nutrient content can vary widely between different plant and animal varieties, with these differences being statistically and nutritionally significant. For example, apricot varieties can provide between less than 1% and more than 200% of the recommended daily intake for vitamin A (Kennedy and Burlingame, 2003). Often, the varieties present in home gardens contain higher nutritional values in terms of essential micronutrients than their commercial counterparts (Kehlenbeck and Maass, 2004). Take the banana, the common variety seen in most stores – the Cavendish variety – has a pro-vitamin A carotenoid content of only about 25 µg per 100 g, though some local banana varieties such as *karat* have more than 8,000 µg per 100 g (Englberger et al., 2003a, 2003b). This diversity maintained in home gardens can help manage the food needs of their families and balance the different micronutrient requirements in their daily diets (Shrestha et al., 2004). It is likely that households do not obtain all of the food sources they need to feed their families but instead supplement these needs with staple crops. In Balinese villages, Sujarwo and Caneva (2015) found that villagers relied on a large range of both local and conventional crops as food sources (see Box 3.2 on Balinese home gardens).

Contribution to well-being

Home gardens go beyond the bounds of nourishing a family; they also support the overall well-being of the household. In the tropics, some estimate that home gardens support nearly 1 billion people (Heywood, 2013). Whether in a rural or urban environment, the biodiverse microenvironment created by home gardens can contribute to a pleasant living environment for families (Pushpakumara et al., 2012). In households of lower socio-economic status, where time and resources are limited, the aesthetic features may not act as an impetus for creating a home garden, however these gardens rich in diversity in both rural and urban communities likely provide numerous mental benefits yet to be measured.

While some gardens serve as a safe haven or refuge, others may have social, cultural or spiritual significance. For example, in the royal gardens of Rana, Nepal, the species diversity serves as a symbol of high social status (Shrestha et al., 2004). In other areas, gardens provide a gathering place for different cultures, as seen within the intercultural community gardens of Germany and Austria (Salako et al., 2014). The convergence of cultures in these gardens encourages a diverse variety of species as influenced from the home gardener's native culture (Schermer, 2014). Culture and cuisine can have a large bearing on the species composition of a home garden (Shrestha et al., 2004). In some gardens, certain species will be cultivated for the purpose of religious or cultural festivities, even further contributing to the conservation of rare and unique species.

Additionally, home gardens provide a wide array of goods and services apart from food and recreation, including fodder, medicines, dyes, fuels and timber (Gautam et al., 2009). For many families in remote areas, health care facilities and resources may be limited. Home herbal gardens, conceived by Foundation for Revitalisation of Local Health Traditions (FRLHT) in India and cultivated by women home gardeners, maintain a diverse array of traditional medicinal species that promote health care access while conserving these important plants (Payyappallimana and Subramanian, 2015). Therefore, plant-based medicines grown in garden plots can leverage local knowledge to maintain the health of individuals in these remote communities (Shrestha et al., 2004).

Contribution to income and sustainable livelihoods

For some households, gardens rich in diversity serve as an essential component to their family's livelihood. Studies of home gardens have shown that households with reasonable market access may supplement their incomes through the sale of animals and crops (Kehlenbeck and Maass, 2004; Shrestha et al., 2004; Sujarwo and Caneva, 2015). Trinh and colleagues (2003) reported the high portion of income from Vietnamese home gardens, where more than 50% of the home garden products are sold. Some markets may welcome diverse plant and animal varieties, thus further encouraging the maintenance of biodiversity in the home gardens (Trinh et al., 2003). Yet there is not always a market demand for traditional varieties. Thus concomitantly with mounting economic and market pressures burdening families, many households are sacrificing the diverse functionality for commercialized modern varieties. This tension between commercialization, conservation and development is a common struggle of home gardeners around the world (Trinh et al., 2003; Abdoellah et al., 2006).

For many households, the home gardens meet two primary needs: food security and income generation. Depending on the household's specific needs, home gardeners may orient the garden primarily to one need or the other, which – depending on the economic conditions in the region – could impact the conservation value of the home gardens (Trinh et al., 2003).

As communities experience economic development, increased physical infrastructure and formalized markets, the number of diverse species in home gardens often decreases (Galluzzi et al., 2010). Birol et al. (2007) found that in Hungary, home gardens in geographically isolated and often marginalized communities often have the greatest species diversity. Therefore, as market pressures continue to face home gardeners, the potential of home gardens to serve as sources of conservation is threatened. Policy efforts can play a role in reducing these pressures, thus preserving and perhaps furthering the role home gardens play in biodiversity conservation.

Contribution to ecosystem services

Diverse home garden agro-ecosystems not only benefit the health and well-being of people, but also the environment. As discussed thus far, home gardens are important to providing provisioning services through the production of food, as well as cultural services. These production systems also contribute to supporting services, which supports all other ecosystem services (e.g., nutrient recycling) and regulating services, which consist of environmental benefits.

Home gardens are often characterized by low-impact methods, including limited to no use of pesticides or other chemicals. These sustainable farming practices also promote species-rich environments that contribute a wealth of ecosystem services in urban and rural areas, including serving as a habitat for pollinators, supporting nutrient recycling, sequestering carbon, providing a refuge for micro and macro-fauna and supporting gene flows between the garden and the surrounding environment (Gautam et al., 2009; Galluzzi et al., 2010). In some instances, the flowering vegetables and herbs growing within managed, diverse home garden habitats can be better resources for the pollinator community than the wild flora found in the surrounding natural habitat (Norfolk et al., 2014). These flowering plants are particularly crucial for communities in hyper-arid desert landscapes, where limited nutrient and water sources limit floral resources in the landscape (Norfolk et al., 2014). In urban areas, home gardens are valued for their contributions to improved air quality, reduced CO_2 emissions and temperatures (Van Veenhuizen, 2006; Viljoen et al., 2009).

Conservation value of home gardens

Home gardens are unique agro-ecosystems that support both in situ and ex situ conservation. In situ conservation occurs through maintenance within natural or even human-made ecosystems. This method promotes the continuous adaptation of plants within the gardens to their surrounding environment. This is a particularly important feature, as opposed to ex situ conservation, or maintaining species in static environments. Home gardeners, based on their knowledge and skills as well as their interest in innovation and experimentation, may favor one conservation technique over the other (Bennett-Lartey et al., 2004). This chapter also provides an additional discussion on the importance of complementary conservation.

Due to in situ practices, many home gardens house a multitude of wild and/or rare species. Formal, ex situ approaches to conservation, such as gene banks, often contain only a small portion of landraces or primitive cultivars, additionally underutilized species and wild relatives are under-represented (Hammer et al., 2003; Galluzzi et al., 2010). A study conducted by Guzman and colleagues (2005) on the genetic diversity of *Capsicum* in Guatemalan home gardens determined that the peppers preserved in the home gardens of the study region were representative of the total diversity of the national *Capsicum* germplasm collection (Guzman et al., 2005).

Complementary in situ and ex situ approaches can be important for germ-plasm conservation in the setting of home gardens, which is to be discussed in greater detail. A recent study by Heraty and Ellstrand (2016) demonstrated that a complementary approach was important to the large variation in genetic diversity within and between populations of maize in home gardens of South-ern California, with migrant farmers playing an important role in fostering and managing this diversity (Heraty and Ellstrand, 2016). Migrant farmers culti-vated traditional maize crops in this new landscape, removed from their centers of diversity (ex situ), while also using traditional practices and their knowledge of the genetic traits as well as management and informal breeding practices (in situ), contributing to the important genetic conservation of maize.

These production systems not only serve as a habitat for species reared and cultivated by the home gardeners, but also for species within the surrounding ecosystem. A study of frugivory bird species of home gardens in Brazil demon-strates how home gardens not only create a habitat for the crops but also sur-rounding species. The study by Goulart et al. (2011) illustrates that agro-forest home gardens provide a diverse, resource-rich habitat for these bird species, providing them food throughout different seasons, thus benefiting the nutri-tional requirements and well-being of these bird species (Guzman et al., 2005).

Whether it is an avian species passing through or a new plant variety grow-ing from its soil, home gardens provide important habitats for genetic diversity. In landscapes consisting of mosaics of forest and agricultural lands, areas of home gardens can act as a refuge or continuum for great landscape connectiv-ity for the likes of amphibians and insects (Pushpakumara et al., 2012). These small-scale production systems, when considered collectively, are profoundly important for conservation efforts worldwide.

Determinants of biodiversity in home gardens

This section highlights four main drivers that impact the inter- and intra-specific diversity of home gardens (Hodel and Gessler, 1999). These factors, as illustrated in Figure 3.1, can act independently and – more often – interact strongly among each other to influence the garden's diversity, thus influencing the value home gardeners reap from gardens to benefit their households (Hodel and Gessler, 1999).

Political factors

This chapter makes the case for effective, evidence-based policies driven by scientific insights to support these biodiverse agricultural production systems. Agricultural and environmental policies have the potential of either promot-ing or constraining inter- and intra-species diversity in home gardens (Casti-ñeiras et al., 2002; Engels, 2002), as the example on policy related to KHGs

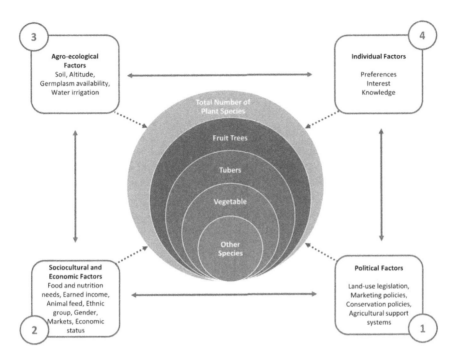

Figure 3.1 Determinants of home garden biodiversity

Source: Adapted from Hodel and Gessler (1999).

in Sri Lanka highlights (see Box 3.3). Appropriate policies of this nature can be further supported by ensuring that home gardens are a major element of any national conservation strategy and policy, and sufficient attention is given to the role and importance of home gardens in agricultural development by extension and research services and agricultural support services in general. Policies and programs may find ways to better support the economic status of families cultivating home gardens through the creation of niche markets. These new market opportunities may increase the demand for diverse species, thus promoting conservation efforts while benefiting farmer livelihoods.

Further work must be done to explore the impact of a nation or region's political actions, or lack thereof, on agricultural biodiversity in home garden production systems and how effective policies can better promote home gardens rich in biodiversity (Eyzaguirre and Bailey, 2009). This action includes the examination of the negative consequences of plant variety protection laws, which can restrict the seed and planting materials available to home gardeners and therefore impact the plant genetic resources present in home gardens (Engels, 2002). In Europe, legislation requires that seeds or planting materials, which are to be exchanged in relatively large quantities, must be registered and deemed uniform, stable and distinct as well as having a proven use value

(Engels, 2002; Eyzaguirre and Bailey, 2009). These requirements set considerable barriers for plants in non-commercialized agricultural systems, thus possibly impacting the flow of germplasm and the plant genetic resources available in the home garden system (Engels, 2002).

National targets, often set in the context of global targets, provide a bellwether for policy makers, moving initiatives to those that benefit plant genetic resource strategies (see Box 3.1). For example, the CBD's Aichi Target 3 aims that

> by 2020, at the latest, incentives, including subsidies, harmful to biodiversity are eliminated, phased out or reformed in order to minimize or avoid negative impacts and positive incentives for the conservation and sustainable use of biodiversity are developed and applied, consistent and in harmony with the Convention and other relevant international obligations, taking into account national socio-economic conditions.

Policies and actions developed in line with targets like this can help to make political factors an effective driver in the promotion and maintenance of biodiversity in home gardens.

Sociocultural and economic factors

Sociocultural and economic factors as alluded to previously in this chapter are an important driver of biodiversity in home gardens. These circumstances include, but are not limited to, geographic proximity to markets, urbanization, migration, policies, traditions/customs, support services (e.g., transport, extension services, distribution of planting material) and economic activities including income generation (Gessler and Hodel, 2004; Salako et al., 2014). Furthermore, a community's social networks are an important contributor to the exchange of information and materials for home gardens, which can stimulate the flow of genes between community households and beyond (Shrestha et al., 2004).

Marginalized social groups that greatly depend on a garden's diversity to support the well-being of their families often are some of the most important contributors to promoting biodiversity within agricultural production systems (Birol et al., 2007; Suwal et al., 2008). These communities that practice home gardening will maximize the contribution of the garden plots to their daily lives while often working with the limited financial and tangible resources available to them.

A household's socioeconomic status is often reflected in the genetic diversity of garden crops and plants. Yongneng and colleagues (2006) found that within home gardens in China there is a significant correlation to household income and species richness (Yongneng et al., 2006). In Nepal, poor households facing more restricted access to land manage less agro-biodiversity than better off households (Adhikari et al., 2004). Home gardeners with more resources and suitable land rights are more willing and able to invest in their gardens and foster a complex, diverse environment (Galluzzi et al., 2010).

In many rural regions, family members migrate to cities in search of off-farm employment to support their families, and consequentially the diversity of home gardens may be threatened. Birol et al. (2005) found that households that experience family members leaving the homestead to find employment elsewhere often rely less on their own home garden products and diversity becomes more simplified (Birol et al., 2005).

Gender dynamics

In many communities, gender differences can be an important factor in creating diverse home garden environments. This influencing factor often varies by location, based on unique sociocultural factors shaping a household (Reyes-García et al., 2010). Often, different genders will place different values on genetic resources based on their traits and uses (Trinh et al., 2003). These values may often be determined based on the role gender plays in the household. Often each gender may play different roles in the upkeep of the home garden, which then influence the garden characteristics and composition (Reyes-García et al., 2010). For example, in some subsistence households of Nepal, women are important caretakers and decision makers because they are responsible for cooking and feeding their families (Shrestha et al., 2004), while men often are responsible for experimenting with and introducing new varieties (Shrestha et al., 2004). When scaling up home garden initiatives, policy makers and planners must keep these unique but inherent dynamics in mind and find appropriate ways of leveraging home gardeners' particular contributions to biodiversity.

Agro-ecological factors

The diversity of plant species found in home gardens is also influenced by the characteristics of the ecological zone in which the garden resides, impacting what home gardeners are able to plant and raise in their gardens. Home gardens often have limited inputs, which further contributes to the adaption of the plant species to their natural environment (Vlkova et al., 2011). For example, home gardens in the savanna zone of Ghana contain less diversity than those in the forest zones as a result of the conditions and resources available in the natural environment (Bennett-Lartey et al., 2004). While ecological factors can affect the structure and diversity of home gardens, limitations may not inhibit a farmer's ability to cultivate a diverse home garden. Trinh et al. (2003) found that in a mountainous region of Vietnam, where the environment is unsuitable for common, commercial cash crops such as fruit and rice, home gardens are important for the subsistence of families (Trinh et al., 2003).

Soil fertility is one such ecological condition that can influence the species or landraces that can be maintained in the home gardens. A study by Junqueira and colleagues (2016) demonstrated that the success of species found in home gardens can in part be attributed to how well-suited the species or landrace is to the soil gradient. Ultimately, the composition of these home gardens is influenced by the home gardener's decisions of what to maintain in the plot. These home gardeners have a great deal of experience and knowledge about the soils and plants, thus knowing the appropriate crops to plant in their gardens' plots.

Individual factors

The home gardener's knowledge, interest and preferences have resonated throughout the discussion about biodiversity found in home gardens thus far. As home gardeners cultivate species, they work to maintain, adapt and disseminate varieties over time that are both suited to the environment and contribute to their household's needs and values. Home gardens promote cultural heritage through the application of traditional agricultural methods and approaches that preserve rare and unique varieties and hence contribute to biodiversity conservation (Birol et al., 2007). Home garden management practices are intricately intertwined in the community's customs and culture. Knowledge and resources for managing these systems are often shared between households, which may help support conservation but also can cause communities to favor some species over others (Eyzaguirre, 2004). While the knowledge of home gardeners is vast, in some regions knowledge gaps may exist regarding the value or diverse functions of a species. For example, some individuals may not be aware of the high nutritional value a crop may hold, which can benefit their family and surrounding communities (Kehlenbeck and Maass, 2004). Thus, it is important to raise awareness through education and events that highlight diversity and the value different varieties can offer to households. In Brazil, with support from the government, garden programs have been incorporated into school curricula to teach children the importance and protection of plant diversity of vegetable and edible wild plants. Helen Keller International's (HKI) Homestead food production (HFP) program also works to raise awareness of diverse household agricultural production systems to benefit nutrition, food security and dietary diversity (Nielsen et al., 2013). The HFP model includes a home gardening program that uses demonstration plots and village model farms to raise awareness and provide trainings on how to cultivate and maintain biodiverse household food systems (Nielsen et al., 2013). As socioeconomic factors – such as globalization and urbanization – continue to reach communities once reliant on traditional home gardens, there are increasing concerns about the erosion of traditional knowledge and awareness and thus the threat to biodiversity conservation.

To combat this erosion, education and outreach efforts, through avenues such as extension services and formal education programs can foster and maintain traditional knowledge, particularly among younger generations. Another solution is to facilitate community biodiversity registers and other community biodiversity management approaches (Figure 3.1), which can be implemented at a low cost (Watson and Eyzaguirre, 2002; Shrestha et al., 2004; Vlkova et al., 2011). These registers can serve as a "working document" available to all community members that maintains information on local cultivated diversity in the community as well as associated traditional ethnobotanical knowledge (Vlkova et al., 2011).

Garden management

A home gardener's management decisions – influenced by their knowledge and experiences – can be highly influential on in situ and ex situ conservation. Home gardeners are faced with myriad tasks and decisions; for example, as unintentional varieties appear within the plots, should they remove these varieties or let them grow to see what benefits the plant may provide? A home gardener's management decisions will dictate the level of biodiversity contained within the garden's bounds. Species selection is dynamic and can depend on a number of on-farm considerations such as perceived value, skills, land/land tenure, materials, capital/credit, labor and biophysical characteristics such as soil, water availability and so forth (Gessler and Hodel, 2004; Galluzzi et al., 2010). The process of obtaining seeds and appropriate planting materials can vary among households based on human or natural selection (Sthapit et al., 2004). Natural selection can benefit underutilized species through the conservation or production of germplasm, if permitted by the home gardener (Sthapit et al., 2004).

Often, the management of home gardens is limited and consists primarily of weeding, irrigating and pruning or cutting down canopy trees, as applicable to the ecosystem (Azurdia and Leiva, 2004). Very often garden litter is left on the soil to decompose (Vlkova et al., 2011). Further, some home gardeners, due to time and interest, may simply not tend to the gardens at all, letting them run their course at the whim of the natural environment, while others will carefully cultivate every corner of their plots (Kehlenbeck and Maass, 2004).

Home gardeners often work with few external inputs such as hired labor or machinery, or the application of synthetic fertilizers or pesticides (Azurdia and Leiva, 2004; Kehlenbeck and Maass, 2004; Vlkova et al., 2011). Instead, home gardeners will use alternative methods for pesticides, such as cutting off infested portions of plants or dusting them with ash (Kehlenbeck and Maass, 2004). Additionally, home gardeners use integrated cropping methods and livestock production to manage the garden's soil fertility (Guzmán et al., 2005). Home garden management approaches make use of the available diversity and resources in the surrounding ecosystem through a harmony between natural and human selection, which shapes the diversity of traditional and modern varieties.

Biodiversity management practices in home gardens

Farmer experimentation and innovation

Human selection is an important component of this equation, as the diversity of a home garden is in part influenced by the curiosity of its gardeners and their interest in experimenting with different varieties (Engels, 2002). Home gardeners may use wild germplasm to create and improve crops (Hughes et al., 2007), while some home gardens may serve as an experimentation plot for new varieties that, if they prove to be beneficial, will be cultivated in larger, commercialized fields (Galluzzi et al., 2010), as home garden crops are physically separated from commercial production. Especially for home gardeners that run farming operations, home gardens provide the freedom to experiment without risking detrimental outcomes on a large-scale production system (Williams, 2004). These sites of experimentation and innovation can serve as a "school" at which young members of the family can learn techniques for domestication of plants and the selection of successful species or traits.

Sites for domestication and diversification

These sites of experimentation and innovation shelter species at various stages of domestication including wild, semi-domesticated and domesticated varieties (Galluzzi et al., 2010). As home gardeners tend to their plots, they contribute to the development and evolution of inter-specific diversity by preserving varieties that demonstrate positive or desired characteristics (Williams, 2004). In some cases, self-sown plants may be tolerated and protected within a plot if they prove to have value (Cruz-Garcia and Struik, 2015). As home gardeners continue to domesticate and diversify crops, they are in consequence developing new germplasm and contributing to the development of the gene pool and management of local biodiversity (Engels, 2002). As domestication activities are influenced by a home gardener's knowledge as well as sociocultural factors, crop domestication can vary widely by region.

Natural forces also contribute to diversification within plots. As there is little spatial area to separate different crop varieties, frequent gene flow facilitates continuous germplasm exchange (Engels, 2002; Galluzzi et al., 2010). Genetic exchanges may occur between cultivated plants and its wild relatives in the surrounding natural area (Galluzzi et al., 2010). Additionally, crops can be cross-pollinated to create hybrid populations. Home gardeners then may detect these populations and decide to propagate them within their plots (Williams, 2004).

Germplasm and seed sharing and seed exchange

Germplasm and seed sharing and exchange occur through natural and human-driven pathways within and beyond the home gardens. Within a village, seeds, seedlings and plant materials may be exchanged between individuals or in a market setting (Vlkova et al., 2011). Additionally, as people travel beyond the village, for example for other off-farm employment, they may acquire new species to introduce (Shrestha et al., 2004) or carry plant material to other locations for propagation. Gardens may also contain varieties from research institutions. An array of natural and anthropogenic activities can dictate the exchange of genetic resources; even neighboring gardeners may have vastly different inter- and intra-specific diversity in their garden plots (Guarino and

Hoogendijk, 2004). Home gardeners exchange not only genetic materials but also knowledge and practices that may promote specific varieties over others. In some cases, select home gardeners, sometimes referred to as "nodal farmers," can help to evolve the informal seed sector through the exchange of information and materials (Vlkova et al., 2011). Whether expanding home garden programs in communities or developing to new regions through existing home garden system, the "nodal farmers" can play a key role in advancing conservation objectives.

Refuge for rare and unique crop varieties and landraces

Due to the home gardener's practice of domestication and diversification of crop species, home gardens are important places for genetic diversity unique and rare in the larger agro-ecosystem (see Box 3.4). Landraces represent such diversity, for these regional varieties of domesticated species are well-adapted to their environment, which make them crucial assets to home gardeners. Landraces can also be nutritionally and culturally important to many communities (Birol et al., 2005). In home gardens, gardeners may cultivate landraces in isolation or conjunction with modern varieties (Galluzzi et al., 2010). Further, indigenous species and rare varieties that are never (or no longer) cultivated in large commercial operations are often harbored in home gardens (Trinh et al., 2003; Galluzzi et al., 2010). This characteristic further solidifies the crucial contribution of home gardens to conservation efforts.

Threats to the future of biodiversity in home gardens

In this period of rapid global change, diverse home gardens may give way to an array of sociocultural and socioeconomic pressures. In regions where gardens are an important source of income, home gardeners may be overcome by market pressures that often encourage the production of cash crops and market-preferred varieties (Vlkova et al., 2011). These pressures cause home gardeners to abandon traditional agronomic practices for new and modern plant varieties, technologies and approaches to cultivation (Azurdia and Leiva, 2004; Shrestha et al., 2004). Urbanization and population pressures further exacerbate this transition to market-based enterprises, encouraging high yielding varieties. Home gardens may also experience fragmentation due to such factors. As families seek off-farm jobs within the city or face time constraints, they may abandon their gardens, leading to biodiversity loss (Reyes-García et al., 2012). Within the cities, urban home gardens are also exposed to space and infrastructure pressures, as well as changing tastes and preferences away from local crops to commercialized products.

Very often, these traditional farming systems do not have support from government agencies to withstand these external forces. In some regions, communities are reclaiming the power of home gardens through the support of non-governmental organizations (NGOs), which promote agro-biodiversity within the gardens through education and technical support (Galluzzi et al., 2010). International and national partners can further help combat these threats to traditional, agro-diverse home gardens by participating in their promotion and scaling up.

Scaling-up species-diverse home gardens: importance of home gardens for complementary conservation

The range of methodologies available for genetic resources conservation can be divided into two strategies: ex situ and in situ. Ex situ conservation refers to the maintenance of plant germplasm away from its natural habitat in facilities such as field gene bank or germplasm centers, whereas in situ refers to maintenance of material in its natural habitat such as farms or home gardens. Although there is general agreement in scientific circles that in situ and ex situ conservation strategies are complementary and one approach cannot replace the other, there has been debate for and against both approaches. Article 9 of the Convention on Biodiversity (CBD) promotes complementary approaches to the conservation of plant germplasm while stressing the particular importance of in situ conservation. Generally, in situ conservation has appeal because it is dynamic and allows for continued adaptation and evolution of the species in its natural habitat compared to the semi-static to static nature of ex situ conservation. When formulating an overall conservation strategy for plant genetic resources, it is vital to incorporate a combination of in situ and ex situ methods for them to complement one another, and home gardens should be an important component of this.

Home gardens, especially in rural areas, tend to contain a wide spectrum of species, such as vegetables, fruits, medicinal plants and spices. They also often contain unique intra-species diversity (see Box 3.4). Home gardens, as single units, may be of little value in terms of conservation, but as the case study of KHGs (Box 3.3) highlights, a community of them in a given area contributes significantly to conservation and genetic diversity. Most of such diversity could in many instances be unique, threatened or rare, as people tend to grow unique materials as well as a broad range of neglected and underutilized species in their gardens. However, the home garden agro-ecosystem is vulnerable to changes in management practices and many other drivers of change. Home gardens are also known to be testing grounds for the farmer or home gardener and a location for testing out some of the wild and semi-wild species – the process of domestication in action. Thus, in rural areas, home gardens will continue to play a role in genetic diversity conservation as well as development.

Home gardens are often overlooked by policy makers as a solution to conservation (Bioversity International, 2013). While research on the value of home gardens for conservation is vast, it is difficult to demonstrate how models for traditional home gardens can be scaled up (Gautam et al., 2009). These small-scale production systems are not highly visible and are often managed by individual households that are not well organized to lobby policy makers

(Gautam et al., 2009). Thus, bottom–up approaches must be conducted to raise the voices of the home gardeners, for these are the individuals who have the depth of knowledge about local agro–biodiversity and local conservation priorities. With the knowledge and support of home gardeners, policy makers and planners can work to link conservation efforts in home gardens to broader urban and rural development efforts (Bioversity International, 2013).

Box 3.4 Home gardens and taro diversity

Root crops are important components of species diverse home gardens and such locations can be an important dynamic conservatory of genetic diversity maintained by rural households. Women are often responsible for maintenance of home gardens and the diversity present is often correlated to multiple uses of plants. Surveys in the Pacific have shown that taro (*Colocasia esculenta*) is among the most dominant plant found in home gardens in Papua New Guinea, Fiji and Tonga. In Asia and the Pacific, different varieties of taro are often maintained in proximity to rural households to meet staple, vegetable and animal food, medicinal and cultural exchange needs. Twenty-four taro landraces were reported in a baseline survey of home gardens in Nepal, with eight being the highest number of landraces maintained by a single household (Rana et al., 2000). Similar surveys in the Philippines recorded 14 taro landraces in home gardens (Pardales et al., 1999). Taro diversity is also important in Vietnamese home gardens where corms, petioles and leaves are used for consumption (Hodel and Gessler, 1999; Trinh et al., 2003). Home gardens represent locations were unique varieties of taro are maintained and as such should be targets for plant collection missions and ex situ conservation. They also need further study as an additional method for in situ conservation. Such information suggests that home gardens have an important role to play in any complementary conservation strategy for maintaining diversity of taro landraces.

Linkages between these efforts not only raises attention to the importance of home gardens in local biodiversity conservation but also supports the development of an interconnected landscape matrix that serves as a better habitat for diverse species, including avian populations (Goulart et al., 2011). In urban areas, "ecological land use complementation" encourages the synergistic interaction of linked habitats to support biodiversity (Colding, 2007).

To support the functioning of these community-based systems, further capacity development support is essential, perhaps through agricultural extension efforts and community mobilization. These efforts must ensure home gardeners have access to the necessary resources and information to maximize the value they can obtain from the gardens while continuing to support diverse genetic resources. These resources may include domesticated priority species, which extension workers can help to integrate into home gardens (Hunter et al., 2015). Support may also include providing home gardeners with the tools and resources to develop markets for home garden products (Bioversity International). With this market development, separate initiatives can advance consumer awareness about the value of home garden products, such as the nutritional benefits of traditional crops. In coordination with capacity development, policy makers must work to ensure farmers have access to appropriate land rights so that they are able to make long-term investments in their gardens that better support conservation efforts as well as the well-being of the household.

It is the diverse values and multiple functions of home gardens which ensures they have much to contribute to complementary conservation approaches, but this is often not well appreciated by population geneticists or conservation biologists because of the small population size of homestead crops and trees. However, it is argued that a network of diverse sets of home gardens in landscapes can serve as part of a meta-population and that unique populations are maintained through the social process of seed/planting materials exchange, natural gene flow and colonization of most preferred and locally adapted seed/planting materials.

There are several practical examples of how home gardeners maintain heirloom chilies, sponge gourds (*Luffa* spp.), tomatoes and fruit trees in home garden systems (Sthapit et al., 2016; Puspakumara et al., 2016). Sthapit and colleagues (2013) have observed that most home garden custodian farmers have a special appreciation of genetic diversity and the ability to identify specific traits and carry out selection of materials for future cultivation (Sthapit et al., 2013). There is much potential for home gardeners as farmer breeders and innovators, local resources and change agents in communities. The home gardens themselves serve as planting material hubs (nurseries) and as platforms for integrating diverse sources of knowledge and skills on fruits and vegetables. Results from a recently completed Global Environment Facility (GEF)-funded Tropical Fruit Tree Diversity project in Asia clearly indicates that that home gardens can maintain rich and special sets of genetic resources that provide global benefits (Sthapit et al., 2016).

One home garden may be inconsequential in conservation efforts, but a community of them with diverse horticultural and other species certainly does offer important benefits and opportunities. Efforts are required to evaluate the role of home gardens (i.e., a community of home gardens) in the conservation of horticultural and permaculture genetic resources, as it is ignored in natural forest genetic resources as well as crop genetic resource conservation efforts. Efforts are needed to explore how home garden systems can be integrated into broader conservation strategies.

Community biodiversity management (CBM) approaches which aim to integrate people, home gardens and production systems encourages the custodianship of land and agricultural biodiversity as a means for improving the livelihoods of local communities and which simultaneously helps maintains important genetic resources and supports evolutionary processes. While the sets of practices implemented from site to site will be context-specific, the principles of CBM can be employed in all contexts. This suggests that the CBM approach has the potential to function as a framework to assist home garden conservation efforts and practices if it is further conceptualized, validated, institutionalized and mainstreamed. The approach is considered a pragmatic method to realize on-site management of local biodiversity and has already been applied by a number of institutions and countries (Sthapit et al., 2012; De Boef et al., 2012, 2013).

In terms of complementary conservation efforts, home gardens play important roles to maintain and safeguard rare, unique and endangered species, varieties and traits. Home gardens can protect such a portfolio of species and varieties as numbers are always small and few. Agricultural biodiversity maintained in home gardens also ensures household members a diversified diet that can lead to improved nutrition and family well-being and provides opportunities for the commercialization of traditional recipes or products which make use of nutritious minor crops and landraces.

Finally, researchers must continue to study home gardens in order to gather data on its economic, nutritional and environmental contributions (Bioversity International, 2013). This evidence can demonstrate to policy makers that home garden initiatives are important and worth investments that hugely benefit the well-being of individuals and the conservation and use of biodiversity.

Acknowledgment

This chapter is dedicated to Dr. Bhuwon Sthapit, who sadly passed away during the development of this book. His dedicated service to conservation and biodiversity will always be remembered.

References

Abdoellah, O.S., Hadikusumah, H.Y., Takeuchi, K. and Okubo, S., 2006. Commercialization of homegardens in an Indonesian village: Vegetation composition and functional changes. In *Tropical homegardens*. Dordrecht, Netherlands: Springer, pp. 233–250.

Adhikari, A., Singh, D., Suwal, R., Shrestha, P. and Gautam, R., 2004. The role of gender in the home garden management and benefit-sharing from home gardens in different production system of Nepal. In R. Gautam, B., Sthapit and P. Shrestha, eds. *Home gardens in Nepal*. Pokhara, Nepal; Rome, Italy; and Kathmandu, Nepal: Local Initiatives for Biodiversity, Research and Development, Bioversity International and Swiss Agency for Development and Cooperation, pp. 84–98.

Azurdia, C. and Leiva, J.M., 2004. Home-garden biodiversity in two contrasting regions of Guatemala. In P. Eyzaguirre and O.F. Linares, eds. *Home gardens and agrobiodiversity*. Washington, DC, USA: Smithsonian Books, pp. 168–184.

Bennett-Lartey, S.O., Ayernor, G.S., Markwei, C.M., Asante, I.K., Abbiw, D.K., Boateng, S.K., Anchirinah, V.M. and Ekpe, P., 2004. Aspects of home-garden cultivation in Ghana:

Regional differences in ecology and society. In P. Eyzaguirre and O.F. Linares, eds. *Home gardens and agrobiodiversity.* Washington, DC, USA: Smithsonian Books, pp. 148–167.

Bernholt, H., Kehlenbeck, K., Gebauer, J. and Buerkert, A., 2009. Plant species richness and diversity in urban and peri-urban gardens of Niamey, Niger. *Agroforestry Systems*, 77(3), pp. 159–179.

Bioversity International, 2013. *Home gardens in Nepal: Impact assessment brief number 10.* Rome, Italy: Bioversity International.

Birol, E., 2004. *Valuing agricultural biodiversity on home gardens in Hungary: An application of stated and revealed preference methods.* Doctoral dissertation, University of London.

Birol, E., Bela, G. and Smale, M., 2005. The role of home gardens in promoting multi-functional agriculture in Hungary. *EuroChoices*, 4(3), pp. 14–21.

Birol, E., Bela, G. and Smale, M., 2007. The role of home gardens in promoting multi-functional agriculture in Hungary. *EuroChoices*, 4(5), pp. 14–21.

Buchmann, C., 2009. Cuban home gardens and their role in social-ecological resilience. *Human Ecology*, 37(6), pp. 705–721.

Caballero-Serrano, V., Onaindia, M., Alday, J.G., Caballero, D., Carrasco, J.C., McLaren, B. and Amigo, J., 2016. Plant diversity and ecosystem services in Amazonian homegardens of Ecuador. *Agriculture, Ecosystems & Environment*, 225, pp. 116–125.

Calvet-Mir, L., Gómez-Baggethun, E. and Reyes-García, V., 2012. Beyond food production: Ecosystem services provided by home gardens: A case study in Vall Fosca, Catalan Pyrenees, Northeastern Spain. *Ecological Economics*, 74, pp. 153–160.

Castiñeiras, L., Mayor, Z.F., Shagarodsky, T., Moreno, V., Barrios, O., Fernández, L. and Cristóbal, R., 2002. Contribution of home gardens to in situ conservation of plant genetic resources in farming systems: Cuban component. In J.W. Watson and P.B. Eyzaguirre, eds. *Home gardens and in situ conservation of plant genetic resources in farming systems: Proceedings of the second international home gardens workshop, Witzenhausen, federal republic of Germany, 17–19 July, 2001.* Rome, Italy: International Plant Genetic Resources Institute, pp. 42–55.

Colding, J., 2007. "Ecological land-use complementation" for building resilience in urban ecosystems. *Landscape and urban planning*, 81(1–2), pp. 46–55.

Convention on Biological Diversity, 2012. *Global strategy for plant conservation: 2011–2020.* Richmond, UK: Botanic Gardens Conservation International.

Cruz-Garcia, G.S. and Struik, P.C., 2015. Spatial and seasonal diversity of wild food plants in home gardens of Northeast Thailand. *Economic Botany*, 69(2), pp. 99–113.

Das, T. and Das, A.K., 2015. Conservation of plant diversity in rural homegardens with cultural and geographical variation in three districts of Barak Valley, Northeast India. *Economic Botany*, 69(1), pp. 57–71.

De Boef, W.S., Subedi, A., Peroni, N., Thijssen, M. and O'Keeffe, E. eds., 2013. *Community biodiversity management: Promoting resilience and the conservation of plant genetic resources.* London, UK: Routledge.

De Boef, W.S., Thijssen, M.H., Shrestha, P., Subedi, A., Feyissa, R., Gezu, G., Canci, A., da Fonseca Ferreira, M.A.J., Dias, T., Swain, S. and Sthapit, B.R., 2012. Moving beyond the dilemma: An assessment of practices contributing to in situ conservation of agrobiodiversity. *Journal of Sustainable Agriculture*, 36, pp. 788–809.

Engels, J., 2002. Home gardens: A genetic resources perspective. In J.W. Watson and P.B. Eyzaguirre, eds. *Home gardens and in situ conservation of plant genetic resources in farming systems: Proceedings of the second international home gardens workshop, Witzenhausen, federal republic of Germany, 17–19 July, 2001.* Rome, Italy: International Plant Genetic Resources Institute, pp. 3–9.

Englberger, L., Aalbersberg, W., Fitzgerald, M.H., Marks, G.C. and Chand, K., 2003a. Provitamin A carotenoid content of different cultivars of edible pandanus fruit. *Journal of Food Composition and Analysis*, 16(2), pp. 237–247.

Englberger, L., Aalbersberg, W., Ravi, P., Bonnin, E., Marks, G.C., Fitzgerald, M.H. and Elymore, J., 2003b. Further analyses on Micronesian banana, taro, breadfruit and other foods for provitamin A carotenoids and minerals. *Journal of Food Composition and Analysis*, 16(2), pp. 219–236.

Eyzaguirre, P., 2004. Foreword. In R. Gautam, B., Sthapit and P. Shrestha, eds. *Home gardens in Nepal*. Pokhara, Nepal, Rome, Italy, and Kathmandu, Nepal: Local Initiatives for Biodiversity, Research and Development, Bioversity International and Swiss Agency for Development and Cooperation, p. iv.

Eyzaguirre, P. and Bailey, A., 2009. International case studies and tropical home gardens projects: Offering lessons for a new research agenda in Europe. In A. Bailey, P. Eyzaguirre and L. Maggioni, eds. *Crop genetic resources in European home gardens: Proceedings of a workshop, 3–4 October, 2007, Ljubljana, Slovenia*. Rome, Italy: Bioversity International, pp. 62–69.

Eyzaguirre, P. and Linares, O.F., 2004. Introduction. In P. Eyzaguirre and O.F. Linares, eds. *Home gardens and agrobiodiversity*. Washington, DC, USA: Smithsonian Books, pp. 1–28.

Freedman, R.L., 2015. Indigenous wild food plants in home gardens: Improving health and income with the assistance of agricultural extension. *International Journal of Agricultural Extension*, 3(1), pp. 63–71.

Gajanana, T.M., Dinesh, M.R., Mysore, S., Sthapit, B., Lamers, H.A.H., Reddy, B.M.C., Rao, V.R. and Dakshinamoorthy, V., 2016. Multi-varietal Orchards: An age-old conservation practice in mango. In B. Sthapit, H.A.H. Lamers, V.R. Rao and A. Bailey, eds. *Tropical fruit tree biodiversity: Good practices for in situ and on-farm conservation*. London, UK: Earthscan and Routledge.

Galluzzi, G., Eyzaguirre, P. and Negri, V., 2010. Home gardens: Neglected hotspots of agrobiodiversity and cultural diversity. *Biodiversity and Conservation*, 19(13), pp. 3635–3654.

Gautam, R., Sthapit, B., Subedi, A., Poudel, D., Shrestha, P. and Eyzaguirre, P., 2009. Home gardens management of key species in Nepal: A way to maximize the use of useful diversity for the well-being of poor farmers. *Plant Genetic Resources*, 7(2), pp. 142–153.

Gessler, M. and Hodel, U., 2004. A case study of key species in Southern Vietnam: Farmer classification and management of agrobiodiversity in home gardens. In P. Eyzaguirre and O.F. Linares, eds. *Home gardens and agrobiodiversity*. Washington, DC, USA: Smithsonian Books, pp. 215–233.

Goulart, F.F., Vandermeer, J., Perfecto, I. and da Matta-Machado, R.P., 2011. Frugivory by five bird species in agroforest home gardens of Pontal do Paranapanema, Brazil. *Agroforestry Systems*, 82(3), pp. 239–246.

Guarino, L. and Hoogendijk, M., 2004. Microenvironments. In P. Eyzaguirre and O.F. Linares, eds. *Home gardens and agrobiodiversity*. Washington, DC, USA: Smithsonian Books, pp. 31–40.

Guzmán, F.A., Ayala, H., Azurdia, C., Duque, M.C. and De Vicente, M.C., 2005. AFLP assessment of genetic diversity of Capsicum genetic resources in Guatemala. *Crop Science*, 45(1), pp. 363–370.

Hammer, K., Arrowsmith, N. and Gladis, T., 2003. Agrobiodiversity with emphasis on plant genetic resources. *Naturwissenschaften*, 90(6), pp. 241–250.

Hemp, A., 2006. The banana forests of Kilimanjaro: Biodiversity and conservation of the Chagga homegardens. *Biodiversity & Conservation*, 15(4), pp. 1193–1217.

Heraty, J.M. and Ellstrand, N.C., 2016. Maize germplasm conservation in Southern California's urban gardens: Introduced diversity beyond ex situ and in situ management. *Economic Botany*, 70(1), pp. 37–48.

Heywood, V., 2013. Overview of agricultural biodiversity and its contribution to nutrition and health. In J. Fanzo, D. Hunter, T. Borelli, and F. Mattei, eds. *Diversifying food and diets: Using agricultural biodiversity to improve nutrition and health*. London, UK: Earthscan, pp. 35–67.

Hodel, U. and Gessler, M., 1999. In situ conservation of plant genetic resources in home gardens of Southern Vietnam. In *A report of home garden surveys in southern Vietnam*. Rome, Italy: SDC and IPGRI, pp. 1–98.

Hughes, C.E., Govindarajulu, R., Robertson, A., Filer, D.L., Harris, S.A. and Bailey, C.D., 2007. Serendipitous backyard hybridization and the origin of crops. *Proceedings of the National Academy of Sciences*, 104(36), pp. 14389–14394.

Hunter, D., Burlingame, B. and Remans, R., 2015. Biodiversity and nutrition. In *WHO/CBD: Connecting global priorities: Biodiversity and human health: A state of knowledge review*. Geneva, Swithzerland: World Health Organization WHO.

Jones, K.M., Specio, S.E., Shrestha, P., Brown, K.H. and Allen, L.H., 2005. Nutrition knowledge and practices and consumption of vitamin a rich plants by rural Nepali participants and nonparticipants in a kitchen-garden program. *Food and Nutrition Bulletin*, 26(2), pp. 198–208.

Junqueira, A.B., Souza, N.B., Stomph, T.J., Almekinders, C.J.M., Clement, C.R. and Struik, P.C., 2016. Soil fertility gradients shape the agrobiodiversity of Amazonian homegardens. *Agriculture, Ecosystems and Environment*, 221, pp. 270–281.

Kahane, R., Hodgkin, T., Jaenicke, H., Hoogendoorn, C., Hermann, M., Keatinge, J.D.H., Hughes, J., Padulosi, S. and Looney, N., 2013. Agrobiodiversity for food security, health and income. *Agronomy for Sustainable Development*, 33, pp. 671–693.

Karyono, I., 1990. Home gardens in Java: Their structure and function. In K. Landauer and M. Brazil, eds. (2004). *Tropical home gardens*. Tokyo, Japan: United Nations University Press, pp. 138–146.

Kehlenbeck, K. and Maass, B.L., 2004. Crop diversity and classification of homegardens in Central Sulawesi, Indonesia. *Agroforestry Systems*, 63(1), pp. 53–62.

Kennedy, G. and Burlingame, B., 2003. Analysis of food composition data on rice from a plant genetic resources perspective. *Food Chemistry*, 80, pp. 589–596.

Kortright, R. and Wakefield, S., 2011. Edible backyards: A qualitative study of household food growing and its contributions to food security. *Agriculture and Human Values*, 28(1), pp. 39–53.

MRSC, n.d. *Livestock and other farm animals*. [Online] Available at: www.mrsc.org/Home/Explore-Topics/Public-Safety/Licensing-and-Regulation/Animal-Control/Livestock-and-Other-Farm-Animals.aspx [Accessed 11 May 2016].

Nielsen, J., Haselow, N., Osei, A. and Zaman, T., 2013. Case study 7: Diversifying diets: Using agricultural biodiversity to improve nutrition and health in Asia. In J. Fanzo, D. Hunter, T. Borelli and F. Mattei, eds. *Diversifying food and diets: Using agricultural biodiversity to improve nutrition and health*. London, UK: Earthscan, pp. 303–312.

Norfolk, O., Eichhorn, M.P. and Gilbert, F., 2014. Culturally valuable minority crops provide a succession of floral resources for flower visitors in traditional orchard gardens. *Biodiversity and Conservation*, 23(13), pp. 3199–3217.

Pardales, J.R., Bañocb, D.M., Yamauchib, A., Lijimab, M. and Konob, Y., 1999. Root system development of cassava and sweet potato during early growth stage as affected by high root zone temperature. *Plant Production Science*, 2(4), pp. 247–251.

Pavia, R., Barbagiovanni, I., Strada, G.D., Piazza, M.G., Engel, P. and Fideghelli, C., 2009. Autochthonous fruit tree germplasm at risk of genetic erosion found in home gardens in the region of Latium (Italy). In A. Bailey, P. Eyzaguirre and L. Maggioni, eds. *Crop genetic resources in European home gardens: Proceedings of a workshop, 3–4 October 2007, Ljubljana, Slovenia.* Rome, Italy: Bioversity International, pp. 21–25.

Payyappallimana, U. and Subramanian, S.M., 2015. Biodiversity and nutrition. In *WHO/ CBD: Connecting global priorities: Biodiversity and human health: A state of knowledge review.* Geneva, Swithzerland: World Health Organization WHO.

Poot-Pool, W.S., van der Wal, H., Flores-Guido, S., Pat-Fernandez, J.M. and Esparza-Olguin, L., 2015. Home garden agrobiodiversity differentiates along a rural-peri-urban gradient in Campeche, Mexico. *Economic Botany*, 69(3), pp. 203–217.

Pushpakumara, D.K.N.G., Heenkenda, H.M.S., Marambe, B., Ranil, R.H.G., Punyawardena, B.V.R., Weerahewa, J., Silva, G.L.L.P., Hunter, D. and Rizvi, J., 2016. Kandyan home gardens A time-tested good practice from Sri Lanka for conserving tropical fruit tree diversity. In B. Sthapit, H. Lamers, V. Ramanatha Rao and A. Bailey, eds. *Tropical fruit tree biodiversity: Good practices for in situ and on-farm conservation.* London, UK: Earthscan and Routledge, pp. 127–146.

Pushpakumara, D.K.N.G., Marambe, B., Silva, G.L.L., Weerahewa, J. and Punyawardena, B.V., 2012. A review of research on homegardens in Sri Lanka: The status, importance and future perspective. *Tropical Agriculturist*, 160, pp. 55–125.

Pushpakumara, D.K.N.G., Wijesekara, A. and Hunter, D.G., 2010. Kandyan homegardens: A promising land management system in Sri Lanka. In C. Belair, K. Ichikawa, B.Y.L. Wong and K.J. Mulongoy, eds. *Sustainable use of biological diversity in socio-ecological production landscapes: Background to the Satoyama initiative for the benefit of biodiversity and human well-being.* Montreal, Canada: Secretariat of the Convention on Biological Diversity, pp. 102–108.

Rana, R.B., Rijal, D., Gauchan, D., Subedi, A., Sthapit, B.R., Upadhyay, M.P. and Jarvis, D.I., 2000. *In situ* crop conservation: Findings of agro-ecology, crop diversity and socio-economic baseline study of Begnas eco-site, Kaski, Nepal. *NP Working Paper No.2/2000.* Kathmandu, Nepal, Pokhara, Nepal, and Rome, Italy: NARC/LI and BIRD/IPGRI.

Reyes-García, V., Aceituno, L., Vila, S., Calvet-Mir, L., Garnatje, T., Jesch, A., Lastra, J.J., Parada, M., Rigat, M., Vallès, J. and Pardo-De-Santayana, M., 2012. Home gardens in three mountain regions of the Iberian Peninsula: Description, motivation for gardening and gross financial benefits. *Journal of Sustainable Agriculture*, 36(2), pp. 249–270.

Reyes-García, V., Vila, S., Aceituno-Mata, L., Calvet-Mir, L., Garnatje, T., Jesch, A., Lastra, J.J., Parada, M., Rigat, M., Vallès, J. and Pardo-de-Santayana, M., 2010. Gendered homegardens: A study in three mountain areas of the Iberian Peninsula. *Economic Botany*, 64(3), pp. 235–247.

Salako, V.K., Fandohan, B., Kassa, B., Assogbadjo, A.E., Idohou, A.F.R., Gbedomon, R.C., Chakeredza, S., Dullo, M.E. and Kakaï, R.G., 2014. Home gardens: An assessment of their biodiversity and potential contribution to conservation of threatened species and crop wild relatives in Benin. *Genetic Resources and Crop Evolution*, 61(2), pp. 313–330.

Schermer, M., 2014. Transnational at home: Intercultural gardens and the social sustainability of cities in Innsbruck, Austria. *Habitat Y Sociedad*, 7, pp. 55–76.

Shrestha, P., Gautam, R., Rana, R.B. and Sthapit, B., 2004. Managing diversity in various ecosystems: Home gardens of Nepal. In P. Eyzaguirre and O.F. Linares, eds. *Home gardens and agrobiodiversity.* Washington, DC, USA: Smithsonian Books, pp. 95–122.

Smith, R.M., Thompson, K., Hodgson, J.G., Warren, P.H. and Gaston, K.J., 2006. Urban domestic gardens (IX): Composition and richness of the vascular plant flora and implications for native biodiversity. *Biological Conservation*, 129(3), pp. 312–322.

Soemarwoto, O., 1987. Home gardens: A traditional agroforestry system with a promising future. In H.A. Steepler and P.K.R. Nair, eds. *Agroforestry: A decade of development.* Nairobi, Kenya: International Centre for Research in Agroforestry, pp. 157–170.

Sthapit, B.R., Lamers, H.A.H. and Rao, R., 2013. Custodian farmers of agricultural biodiversity: Selected profiles from South and South East Asia. In *Proceedings of the workshop on custodian farmers of agricultural biodiversity, 11–12 February 2013.* New Delhi, India: Biodiversity International.

Sthapit, B.R., Lamers, H.A.H., Rao, R. and Bailey, A. eds., 2016. *Tropical fruit tree biodiversity: Good practices for in situ and on-farm conservation.* London, UK: Earthscan and Routledge.

Sthapit, B.R., Rana, R.B., Hue, N.N. and Rijal, D., 2004. The diversity of Taro and Sponge Gourd in home gardens of Nepal and Vietnam. In P. Eyzaguirre and O.F. Linares eds. (2009). *Home gardens and agrobiodiversity.* Washington, DC, USA: Smithsonian Books, pp. 234–255.

Sthapit, B.R., Subedi, A.H.A., Lamers, H., Jarvis, D., Rao, R. and Reddy, B.M.C., 2012. Community based approach to on-farm conservation and sustainable use of agricultural biodiversity in Asia. *Indian Journal of Plant Genetic Resources,* 25(1), pp. 88–104.

Sujarwo, W., Arinasa, I.B.K., Caneva, G. and Guarrera, P.M., 2016. Traditional knowledge of wild and semi-wild edible plants used in Bali (Indonesia) to maintain biological and cultural diversity. *Plant Biosystems: An International Journal Dealing with All Aspects of Plant Biology,* 150(5), pp. 971–976. DOI: 10.1080/11263504.2014.994577

Sujarwo, W., Arinasa, I.B.K., Salomone, F., Caneva, G. and Fattorini, S., 2014. Cultural erosion of balinese indigenous knowledge of food and nutraceutical plants. *Economic Botany,* 68(4), pp. 426–437.

Sujarwo, W. and Caneva, G., 2015. Ethnobotanical study of cultivated plants in home gardens of traditional villages in Bali (Indonesia). *Human Ecology,* 43(5), pp. 769–778.

Sunwar, S., Thornström, C.G., Subedi, A. and Bystrom, M., 2006. Home gardens in western Nepal: Opportunities and challenges for on-farm management of agrobiodiversity. *Biodiversity and Conservation,* 15(13), pp. 4211–4238.

Suwal, R., Regmi, B.R., Sthapit, B. and Shrestha, A., 2008. Home gardens are within reach of marginalised people'. *Leisa Magazine,* 24, p. 34.

Taylor, J.R. and Lovell, S.T., 2014. Urban home food gardens in the Global North: Research traditions and future directions. *Agriculture and Human Values,* 31(2), pp. 285–305.

Trinh, L.N., Watson, J.W., Hue, N.N., De, N.N., Minh, N.V., Chu, B.R. Sthapit, B. and Eyzaguirre, P.B., 2003. Agrobiodiversity conservation and development in Vietnamese home gardens. *Agriculture, Ecosystems and Environment,* 97(1–3), pp. 317–344.

United Nations, undated. Transforming our world: The 2030 agenda for sustainable development. *Sustainable Development Knowledge Platform.* [Online] Available at: https://sustainable development.un.org/post2015/transformingourworld [Accessed 12 May 2016].

United Nations Environment Programme and Convention on Biological Diversity, 2010. Decision X/2: The strategic plan for biodiversity 2011–2020 and the Aichi biodiversity targets. *Conference of the Parties to the Convention of Biological Diversity.* Nagoya, Tokyo and Montreal, Canada: Convention on Biological Diversity.

Van Veenhuizen, R., 2006. Introduction. In R. Van Veenhuizen, ed. *Cities farming for the future-urban agriculture for green and productive cities.* Manila, Philippines: RUAF Foundation, International Development Research Centre and International Institute of Rural Reconstruction, pp. 1–17.

Viljoen, A., Bohn, K., Tomkins, M. and Al, E., 2009. Places for people, places for plants: Evolving thoughts on continuous productive urban landscapes. *Proceedings of the 2nd International Conference on Landscape and Urban Horticulture.* Bologna, Italy, 9–13 June. Leuven, Belgium: International Society for Horticultural Science, p. 38.

Vlkova, M., Polesny, Z., Verner, V., Banout, J., Dvorak, M., Havlik, J., Lojka, B., Ehl, P. and Krausova, J., 2011. Ethnobotanical knowledge and agrobiodiversity in subsistence farming: Case study of home gardens in Phong My commune, central Vietnam. *Genetic Resources and Crop Evolution*, 58(5), pp. 629–644.

Watson, J.W. and Eyzaguirre, P.B., 2002. Home gardens and in situ conservation of plant genetic resources in farming systems. *Proceedings of the 2nd International Home Gardens Workshop*. Witzenhausen, Germany, 17–19 July. Rome, Italy: International Plant Genetic Resources Institute.

Weerahewa, J., Pushpakumara, G., Silva, P., Daulagala, C., Punyawardena, R., Premalal, S., Miah, G., Roy, J., Jana, S. and Marambe, B., 2012. Are homegarden ecosystems resilient to climate change? An analysis of the adaptation strategies of homegardeners in Sri Lanka. *APN Science Bulletin*, 2, pp. 22–27.

Williams, D.E., 2004. The conservation and evolution of landraces of peanuts and peppers. In P. Eyzaguirre and O.F. Linares, eds. *Home gardens and agrobiodiversity*. Washington, DC, USA: Smithsonian Books, pp. 256–265.

Yongneng, F., Huijun, G., Aiguo, C. and Jinyun, C., 2006. Household differentiation and on-farm conservation of biodiversity by indigenous households in Xishuangbanna, China. *Biodiversity and Conservation*, 15(8), pp. 2687–2703.

4 Gender and home gardens

Toward food security and women's empowerment

Katarina Huss, D. Hashini Galhena Dissanayake and Linda Racioppi

Introduction

Women's contributions to household food and nutritional security are often undermined in academic and policy discourse. Generally, it is presumed that large-, medium- and smallholder farms are the main sources of food. Such farms are often under the supervision and management, actual or nominal, of male members of the household. As the data in Figure 4.1 demonstrate, women's participation in agriculture is highly variable globally from a low of less than 1% in North America to a high of 59% in South Asia (World Bank, 2019). According to the Food and Agricultural Organization (FAO), about 43% of the agricultural labor force in developing countries consists of women (Raney et al., 2011). These figures vary across regions of the world. Estimated aggregate averages range between 24% and 60% across low- and middle-income countries in Asia; the average is about 55% for countries in sub-Saharan Africa. Yet, there are differences among countries. In South Asia, female employment in agriculture in Nepal is nearly 80% of the total female employment, but it is only about 2% in the Maldives. Similarly, in Africa, nearly everyone in the female workforce in Burundi is employed in agriculture (96%), but in South Africa it is less than 4%. There are also differences with regards to women's contribution to farm activities. Ali (2005) reports that women do more than 80% of agricultural tasks on smallholder land in Bangladesh.

Women's work in agriculture systems including home gardens has become more prevalent as the global economy shifts in developing countries, forcing men to seek employment in urban areas. While it is difficult to accurately determine the extent of agricultural production attributed to women and men, it is often assumed in scholarly and policy making circles that men are the main contributors to agricultural production. Part of the problem is the fact that men are overwhelmingly the holders of agricultural land, leading to the presumption that they are chiefly responsible for agricultural activities. However, women undertake a variety of agricultural roles, from landholders to home gardeners, that contribute to local food systems.

In the absence of men, women have taken up additional responsibilities and non-traditional roles within their households. Some studies report that this

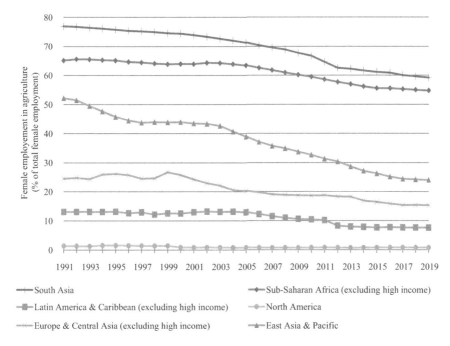

Figure 4.1 Women's participation in agriculture as a percentage of total female employment
Source: World Bank (2019), compiled by authors.

change has given women more control over household affairs and decision-making (Horenstein, 1989; Adekunle et al., 2013; Vazquez-Garcia, 2008). Unfortunately, too often their contributions tend to go unrecognized by other household members and agricultural policy and decision makers. As a result, women face limitations that avert their ability to be more productive and to make autonomous decisions. Only 5% of the global agricultural landholdings are accessible to women as landowners or managers (Food and Agriculture Organization of the United Nations, 2011). The Consultative Group for International Agricultural Research (CGIAR) points out that if women had access to the same resources as men, their on-farm yields would rise by 20% to 30%. Interestingly, home gardens, which are primarily controlled by women, tend to be highly complex, productive and genetically diverse cropping systems (Howard-Borjas and Cuijpers, 2009; Schreinemachers et al., 2015). Home gardens and women's work in home gardens is important to agriculture, development, family nutrition, biodiversity and communities globally. For example, kitchen gardens in Tajikistan, where women have near complete autonomy and authority, have historically contributed a significant proportion to the local food supply, making them a key source for the country's food security (Rowe,

2009). Moreover, within home gardens women conserve, preserve and domesticate flora and fauna species through selection, collection and management (Howard-Borjas and Cuijpers, 2009). While between countries and across cultural contexts there is significant variation in the organization and composition of home gardens and their practices, across contexts gender remains an important consideration for evaluating home gardens. This chapter highlights the vital role of women as key managers of home gardens and discuss their contributions to household food security, household income generation and environmental conservation. It also presents home gardens as grounds for empowering women and enhancing their socioeconomic status. The chapter concludes with recommendations to support women and their home gardening activities.

Women, home gardens and food security

Historically speaking, the use of home gardens to provide food security should not be surprising. Even in more developed countries, during wartime home gardens have been used to combat food scarcity and have relied on women for the work. During World Wars I and II, the UK and the US both involved women in agricultural production and encouraged women to plant home gardens. Americans especially saw a massive increase in the number of home gardens, and the practice of women gardening for their family became so culturally significant that following the war, advertisements for real estate included images of women working in their gardens to provide for their families. However, the US did not maintain their home gardens into the Cold War, despite the US Department of Agriculture's (USDA) attempts at propaganda to encourage home gardens (Gowdy-Wygant, 2013).

The FAO (2001) defines food security as "a situation that exists when all people, at all times, have physical, social and economic access to sufficient, safe and nutritious food that meets their dietary needs and food preferences for an active and healthy life." This conceptualization of food security relies on four conditions: availability, accessibility, utilization and stability. The availability dimension of food security takes into consideration that adequate quantities of quality food are available at all times. Home gardens primarily managed by women contain an assortment of annual and peri-annual crops and livestock that can provide a supplemental source of food throughout the year (Mattsson et al., 2018). Through this function, home gardens help reduce the households' vulnerability to food shortages and improve resilience to adverse shocks. Households in Tajikistan utilized home gardens to supplement household needs during the Cold War. Even after the Cold War, families continue to use and depend upon the land due to the limited economic opportunities that allow women and men to find other viable alternatives (Rowe, 2009). The collapse of the Soviet Union following the Cold War placed similar stress on Russia. Women in Russia were faced with high unemployment rates, a growing income gap, more prevalent cases of sexual harassment and high alcoholism among their husbands, which tasked women with primary care of the household. Home gardens in urban spaces became a method of survival for women. The gardens were so pervasive that even as late as in 2008, when the economy fell, women turned to home gardens to provide for families (Wegren et al., 2017). Case studies from other parts of the world reveal that home gardens are widely used by women to combat household food insecurity and supplement family diets (Horenstein, 1989; Keys, 1999; Ahmad et al., 2007; Girard et al., 2012; Galhena et al., 2013; Marambe et al., 2018).

The literature focusing on access to food takes multiple angles. It entails physical and economic means to access to food stocks available through domestic production or international procurement as well as political/legal provisions and social entitlements to acquire food. Home gardening gives women direct physical access to food. The low cost of production through resource management and reuse and spatial proximity makes food more affordable for the household. As the main caretakers of the home gardens, women have freedom to raise crops and animals that match their culinary and consumption preferences as well as those suitable for post-harvest processing, storage and marketing (Howard-Borjas and Cuijpers, 2009). In some contexts, women's access to food is constrained by prevailing social norms. In most patriarchal culture, male members access to food is prioritized over women and children (Olum et al., 2018; Alonso et al., 2018). While women experience more autonomy, control and opportunities for decision-making with regard to their home gardening activities, intensive efforts are still needed to change deeply rooted cultural norms, attitudes and mindsets that will enable better social access to food and equity for women.

Utilization or absorption is the component of food security that concentrates on food safety, nutrition and health parameters that must be in place for an active and healthy life. Home gardens in tropical areas can account for 3% to 44% of a family's total caloric intake (Kumar and Nair, 2004, p. 141). Men tend to use home gardens to grow small amounts of cash crops or test new plants for their viability as new crops on smallholder land (Howard, 2003; Howard, 2006). Whereas men may be growing highly commercialized crops in home gardens, women tend to focus on plants most likely to benefit family nutrition, health and well-being and only sell their produce when they have excess (Trinh et al., 2003; Ngome and Foeken, 2012). Especially as agricultural smallholders feel pressure from the external economy to shift their crops to cash crops as opposed to food crops, home gardens become important sources of nutrition for families (Horenstein, 1989). Women may dominate home garden work in contexts where there are low education levels and employment outcomes for women and where women are perceived as being responsible for household sustenance (Ngome and Foeken, 2012). As such, women make a positive contribution to their families' nutrition and health through their home gardens. For instance, Helen Keller International's project in Bangladesh helped women develop better home gardening techniques and promoted the cultivation of plants rich in vitamin A to address nutritional deficiencies that cause blindness (Bushamuka et al., 2005). In Kenya, families with the most nourished children were the ones able to devote the most amount of time to home gardening (Horenstein, 1989). As women are the primary child caregivers, their empowerment is likely to improve nutritional and health outcomes (van den Bold et al., 2013).

Women, home gardens and income generation

Home gardens may also relieve some aspects of poverty and can be perceived as a beneficial strategy for poverty alleviation. The food products available through home gardens increase household food supply and reduce its dependence on purchases from the market (Faber et al., 2002) and (Rammohan et al., 2019). The money saved can then be used for purchase of additional food items or for other household purposes. Home gardens can create a profit if they produce surplus. Small amounts of cash crops can also be planted in home gardens. Through the sale of home garden products, households can expand their purchasing power (Girard et al., 2012). A survey from South Africa revealed that 60% of surveyed households relied on their own production for subsistence. If they did sell excess, that money was then used to purchase other foodstuffs (Adekunle et al., 2013).

Across contexts, however, men tend to control the more economically lucrative aspects of home gardens. In Mayan contexts, men tended to be in charge of crops only when they were cash crops or had a high commercial value (Howard, 2006). In Alta Verapaz, Cuba, men and women were equally involved in home gardens, but men tended to be more involved with cash crops (Watson and Eyzaguirre, 2001). Men also tend to sell the excess women grow in the household. In Bangladesh, young married women are not allowed to sell their produce at the market. Thus the income from selling home garden products rarely comes back to benefit them or the family (Wilson-Moore, 1989). When women can sell their excess, the revenues earned impact the household more positively, and women feel as though they made a important contribution to the household. Women in Tamil Nadu, India, selling their medicinal plants expressed that they felt more recognized and accepted in their families because their plants had contributed to social status of the family (Torri, 2012). In Ethiopia, women manage the home gardens and sell the excess on the markets, which gives them more power in the household (Watson and Eyzaguirre, 2001).

However, women may struggle to access all of these benefits due to their limited property ownership and lack of access to resources like arable land and water. The fact that women in many countries do not own their home garden lands may mean that ultimately women could lose access to their livelihood and work in the case of separation or divorce from a husband (Howard, 2006). In addition, limited access to productive resources, seasonality, climate variability and knowledge of best practices and coping mechanism individually or collectively can impact the year-round productivity and profitability of home gardening (Patalagsa et al., 2015; Marambe et al., 2018).

Women, home gardens, biodiversity conservation and preservation of indigenous knowledge

Home gardens are important spaces for biological preservation, and women play an important role in this process. Women preserve biodiversity through growing indigenous plants, diverse cultivars of crops and plants for both medicinal and culinary purposes in their home gardens. Home gardens are often overlooked as a site of conservation and biodiversity because they are so small (Chambers and Momsen, 2007). Home gardens contain a variety of indigenous and diverse cultivars, even those that may not be regarded by outsiders as useful. Women cultivate a variety of plants because women use home gardens to provide for the nutritional and medicinal needs of their family (Watson and Eyzaguirre, 2001; Howard-Borjas and Cuijpers, 2009). It is thus important that women in home gardens preserve culturally significant crops for culinary use and cultivate healthy crops and strong cultivars in order to serve household needs (Howard-Borjas and Cuijpers, 2009).

Since home gardens are constructed to be extensions of the home, women are tasked with the preservation of plants in home gardens (Aguilar-Støen et al., 2009; Howard, 2003). Households managed by women have significantly higher diversity in terms of cultivated plant species than gardens managed by men (Ban and Coomes, 2004). Women in home gardens grow plants that are useful for the household. This incentivizes them to grow a variety of crops that are useful for medicinal purposes and consumption, whereas men may be using home gardens to test agricultural crops not necessarily for production of plants in the garden. In Latin America, most home garden plant species belong to women; men are involved in the home gardens to test plants for agriculture (Howard, 2006; Bain, 1993). In Nepal, men also tend to introduce wild or exotic plants and trees into home gardens that can be used for profit or grown agriculturally, whereas women prefer to grown traditional plants in home gardens (Watson and Eyzaguirre, 2001). Women may also introduce wild plants into their home gardens while testing crop varieties (Howard-Borjas and Cuijpers, 2009). Women in communities tend to share knowledge about a variety of exotic plants and cultivars for home gardens (Bain, 1993; Ban and Coomes, 2004; Márquez and Schwartz, 2008). The practice of introducing exotic plants intended to remain in the home garden contributes to the genetic complexity and diverse cultivars in women's home gardens (Howard-Borjas and Cuijpers, 2009).

Women preserving actual diversity in their gardens is equally important to the indigenous and cultural knowledge that home gardens help women preserve. Women's domestic tasks such as gardening, gathering, cooking and post-harvest preservation are sites of biodiversity conservation that demand a high level of technical knowledge about plants and their uses (Howard, 2003). Women in Africa grow plants in their home gardens that are not traditionally recognized as being useful for consumption. However, women use the plants in traditional cooking, so they continue to preserve the plants (Howard-Borjas and Cuijpers, 2009). As opposed to the economically focused activity of men in home gardens, women may also maintain home gardens for aesthetic purposes. In the West Italian Alps, this had positive impacts on diversity in home gardens too (Mattalia et al., 2018).

Home gardens do not only contain plants for consumption or aesthetics; they often contain medicinal plants as well (Agelet et al., 2000; Augustino and Gillah, 2005; Bain, 1993; Horenstein, 1989; Torri, 2012). Women grow medicinal plants for the same reasons they grow crops for consumption: the medicinal plants supplement a household need. In Tamil Nadu, women prefer medicinal plants to pharmaceuticals because the plants are a low-cost alternative. The medicinal plants are also another source of income that women can use to buy food (Torri, 2012). Women are not exclusively responsible for medicinal plants in home gardens, however. In Tanzania, both men and women are herbalists who work with medicinal plants in the home garden space; men typically work with roots while women cut bark, which could be due to the labor intensity of each action (Augustino and Gillah, 2005). Medicinal plants in home gardens are equally important to biological diversity because women are still cultivating plants based on knowledge of the usefulness of indigenous plants (Howard-Borjas and Cuijpers, 2009).

Women's agricultural knowledge and preference for culinarily and nutritionally useful plants in home gardens can also impact biodiversity in agricultural crops if they are involved in seed selection and seed storage. Seed storage is meant to ensure the health of future crop harvests (Howard-Borjas and Cuijpers, 2009). Women are often in charge of agricultural seed storage if they are involved in post-harvest activities (Howard, 2003). In the case of Bangladesh, the seeds for home gardens and agricultural crops are stored by women in collaboration with their husbands (Oakley and Momsen, 2005). Women choose to preserve different seeds than men. In Oaxaca, Mexico, women were involved in the storing of seeds, but while men generally chose seeds based on resistance to disease, size, adaptation and productivity, women were more like to choose seeds based on taste and texture as they played a role in their cooking preparation (Aguilar-Støen et al., 2009). This mirrors women's choices in home gardens to cultivate plants useful for culinary preparation. Men may also choose seeds for taste or texture if their agricultural produce is processed or used in culinary traditions, though not necessarily for cash crops (Howard-Borjas and Cuijpers, 2009). Women may also be more knowledgeable about the types of seeds most likely to resist climate change or other adversity in the growing seasons. In the case of the Hleketani Community Garden in South Africa, this directly impacted agriculture. Women had stockpiled home garden seeds most likely to survive drought. During an extreme drought season, these seeds were planted in family farms with indigenous plants, and despite the extreme conditions, the crops survived. Women's seed preservation in home gardens and knowledge protected crops from the most extreme effects of drought (Vibert, 2016). Through such activities, women intentionally or unintentionally contribute to preservation of germplasm and progression of indigenous knowledge on home gardening.

Home gardens as spaces for women's empowerment and social engagement

It should be noted that women may not always maintain home gardens for food security, income generation, or conservation. Women have contributed to agriculture in spaces beyond home gardens. In Uttarakhand, India, women manage home gardens, but 78% of women in regular work are employed in agriculture and their work has contributed to the economic and biological diversity of the region (Bargali, 2015). In places where women are not employed or recognized in mainstream agriculture, women may be relegated to home gardening. Home gardens in cultures where women have limited mobility outside the home allow women to contribute the household without requiring them to challenge dominant cultural norms (Vazquez-Garcia, 2008). Women's contributions to the family via home gardens are considered respectable because they are natural extensions of the domestic duties and cultural roles that women play (Howard, 2006; Ngome and Foeken, 2012).

Home gardens being constructed as a feminine space where women have authority has been good for women's power, however. Nicaragua, home gardens are in domestic spaces adjacent to homes on patios; men maintain them physically, but they are in such proximity to the home space that they are considered female spaces in the home (Shillington, 2008). The ownership of the home garden as a home space is pervasive. In Mesoamerica, women appear to have informal rights to home gardens despite men predominately owning the land. The space becomes gendered: the home garden is a space allocated to women (Howard, 2006). In Mayan culture, even women's husbands may have no right to harvest or destroy home garden plants, which highlights the power and social norms about property and land ownership offered by home gardening (Howard, 2006). Women may also assert power through home gardens. For Palestinian women in the 1920s, the Urban Gardens Project followed the model of the World War I US victory gardens. It helped Palestinian women to have a place of their own to meaningfully contribute to the household and "cast roots" to the homeland (Alon-Mozes, 2007).

Home gardens can increase women's power in the household and local community and give women access to additional resources to care for their family. The positive effects of home gardens on women's power and familial nutrition have led some nonprofits and governments to view the home garden as a useful tool in development. It is difficult to measure the actual effect of home gardens on empowerment, however. Women's empowerment is not universally measured. It may be measured in terms of decision-making, land ownership or marriage. Though the positive interaction between empowerment and improved nutrition is known, there is little knowledge about how it occurs (van den Bold et al., 2013). In the case of home gardens, women in Bangladesh cite that they felt their contribution to household income through home gardens granted them more respect from other family members (Bushamuka et al., 2005). This authority may lend itself to women having a greater say in their family in ways that positively impact households and communities (Howard, 2006). However, women feeling empowered because of contributions to the household should be contrasted with women being overlooked in agricultural spaces, limited control over land and profits from home gardens and power outside the household (Keys, 1999).

In Oaxaca, Mexico, women maintained home gardens for aesthetic and social reasons (Aguilar-Støen et al., 2009). Home gardens do not only have to be for income or supplementing diet; they can also be symbols of women's place in society. In Guatemala, women may plant flowers in their gardens alongside medicinal plants, foodstuffs and for aesthetic purposes (Márquez and Schwartz, 2008). In the Brazilian Amazon, the planting of flowers in home gardens is an expression of femininity and household status (Murrieta and WrinklerPrins, 2003). Women in the Ecuadorian highlands may even derive a sense of freedom from the value of her garden, as it represents her work and investment (Finerman and Sackett, 2003). As witnessed in the Brazilian Amazon by Murrieta and WinklerPrins (2003), women's knowledge of gardening may give them authority to challenge the agricultural practices of their husbands.

Home gardens are sites of continuous accrued knowledge of communities (Kumar and Nair, 2004). Knowledge of home gardening techniques and practices is passed along to family members and maintained by the community. Women accumulate traditional knowledge of plants, foods and household projects and use their skills to develop the plants important for their family's needs (Howard-Borjas and Cuijpers, 2009). Women, especially older women, are most often responsible for the transmission of knowledge and considered home gardening specialists. Knowledge and its transmission are thus linked to gender (Howard, 2006). Women's indigenous knowledge should not be overlooked as a valuable site for in situ conservation of crop diversity (Chambers and Momsen, 2007). Women in Tamil Nadu knew from the body of knowledge they had acquired since childhood how to use medicinal herbs, and when they didn't, they had a network of family and friends to ask (Torri, 2012). In households in Mexico and Nicaragua, gardening may even be a shared family activity involving parents and children; this same pattern exists in the Iberian Peninsula (Reyes-García et al., 2010). The gardening techniques not only impart agricultural but also cultural knowledge. Kaqchikel women and elderly relatives teach children about human–environment relationships through prayer and story linked to the food grown in home gardens. Children then acquire home gardening skills and knowledge about indigenous plants and their uses through observation and practice alongside their mothers and other older family members.

Nontraditional home gardens in urban and other spaces

Home gardening is not exclusive to rural areas or developing countries. However, gardening may be less common in urban spaces. Bhatti and Church (2004) find that in the UK, individuals who owned houses are far more likely to have access to gardens; however, many adults viewed gardening as a chore rather than an embodied experience with nature, suggesting that gardening was not a gratifying process. As a result, people chose to garden less (Bhatti and Church, 2004).

Instead of home gardening, community gardening is more common in urban spaces. Community gardens may serve the same function as home gardening rural areas: they are sources of food for household as well as potential spaces to build social networks. Households that participate in community gardens in the US consume fruits and vegetables in greater quantities than their non-gardening neighbors (Blaine et al., 2010; Alaimo et al., 2008). This is particularly important for food security and nutrition in the US for low-income Hispanic and African American communities with limited access to supermarkets and produce (Alaimo et al., 2008; Waliczeck et al., 1996; Gray et al., 2014). Community gardens may even make community members feel more connected with their neighbors, as Gray et al. (2014) found in their study of a new community garden project in San Jose, California. Most research on community gardens studies the rise of community gardens in urban spaces in the US (Guitart et al., 2012). However, urban gardening does occur in other countries too. Women in the Hleketani Community Garden in South Africa use the garden as a space where women from the community come together to grow food, which contributes to poverty reduction and health promotion (Vibert, 2016).

For immigrants or displaced populations, community gardens are places for people to supplement their diets, grow traditional food and preserve cultural practices. In refugee camps, Somali Banut women use community gardens to supplement the dietary needs of their families and use the gardens as spaces for community cooperation (Coughlan and Hermes, 2016). African refugees and immigrants in the US have also planted and invested in individual plots of land in community gardens to grow traditional foods and other crops to supplement their diet. In this case, the community building from community gardens was significantly less important (Harris et al., 2014). Asian immigrants in the US have used gardens as religious or cultural spaces that allow them to connect with their cultural heritage and history (Mazumdar and Mazumdar, 2012).

Community gardens are not always a female-dominated space, however. Buckingham (2005) found that there has been a shift in Britain after World War II to view gardening as a masculine practice enforced by women's perception that preparing the land is too difficult. This point is reinforced by findings from a study of community gardens in St. Louis, Missouri: while in principle men and women had equal opportunity to garden, some tasks were delegated to men based on their presumed physical ability (Parry et al., 2005). A study of community gardening in Rock Island, Illinois, shows that men are the primary gardeners across the garden, but the African refugee and African American gardens are managed by women (Strunk and Richardson, 2019). This suggests that the gendered nature of gardening varies across cultural contexts and depends not only on men's involvement in agriculture but also shifting definitions of masculine and feminine roles.

By way of conclusion: home gardens to enhance food security and gender equity

As discussed earlier, home gardens have the capacity to increase the availability of food as well as households' ability to access and utilize fresh, healthy and nutritious food. The potential for enhanced food security is made possible both through actual garden production and through income generation, which can enable households to purchase foods for a more diverse diet. While more research is needed to establish definitive links between home gardening and improved nutrition, emerging literature suggests that development of home gardens can lead to positive nutritional outcomes. In particular, some studies indicate that home gardens have a positive relationship to women and children's health because the food grown in home gardens tends to be nutrient rich. For instance, an early review by Bain (1993) in Mexico argued that rural women use home gardens to provide nutrient-rich food to their families. In a planned home gardening intervention in South Africa, Faber et al. (2002) find that home gardening improved children's access to vitamin A, an important nutrient for their growth and development. Similarly, investigation of a program in Bangladesh aimed at increasing home gardens demonstrated that home gardening played an important role in combatting Vitamin A and other micronutrient deficiencies (Bushamuka et al., 2005). Schreinemachers et al. (2015) echo this general point, arguing that home gardens in Bangladesh are a successful intervention to increase range of vegetables in rural households, thus contributing to nutrition security. And Rubaihayo (2002) also stresses that in Uganda home gardens growing indigenous vegetables provide an important source of nutrients and protein, important for children and women. Interestingly, the literature also points to the potential of home gardening to augment incomes (Márquez and Schwartz, 2008; Méndez et al., 2001; Bushamuka et al., 2005; Vasquez-Garcia, 2008) and encourage women's empowerment.

Since home gardens are spaces that not only contribute to household food security but also to women's empowerment, there is some attention by academics and development agencies in teaching women to home garden. Women can be more productive in their home gardens with training and policy reform. Ahmad et al. argue women should be provided provide training, proper credit and marketing and irrigation (2007, p. 1176). The outcome of the Helen Keller International program in Bangladesh was that 85% of women felt that they had increased their contribution to the household, as compared to 52% of women in the control group (Bushamuka et al., 2005, p. 21). According to Patalagsa et al. (2015), that program provided women with training on gardening tasks and crop selection and management; however, it did not extend to marketing of production. In Pakistan, training women to engage in home gardening has involved teaching them best practices in pesticide use, management of home gardens, plant selection and explaining the role of water, sunlight and weeding on plants (Yasmin et al., 2013). Some home gardening training for

Pakistani women has also focused on organic cultivation and integrated pest management (Bajwa et al., 2015). Other projects have taught Pakistani women to grow off-season vegetables to improve food security, nutrition, income and employment (Yasmin et al., 2014) and to provide information and technologies to tribal women in India (Chauhan, 2012).

What these studies demonstrate is that training women in home gardening not only gives them access to information and technologies that will improve yields and better nutrition, but it also boosts their confidence and empowers their decision-making (Bajwa et al., 2015). Yet there are challenges to expanding women's participation in more intensive home gardening. For example, training alone cannot provide women with access to necessary inputs such as credit or fertilizer or reliable water supply; neither can it ensure women entry to markets, particularly when sociocultural norms dictate against their market participation. As important as training can be, it is also important to have policies and social practices that support women's income generation and empowerment. Thus when considering interventions to support home gardening activities, it is necessary to recognize the intrinsic value women place on their home gardens and understand the constraints they face.

References

Adekunle, O., Monde, N., Agholor, I. and Odeyemi, A., 2013. The role of home gardens in household food security in Eastern Cape: A case study of three villages in Nkonkobe municipality. *Journal of Agricultural Science*, 6. https://doi.org/10.5539/jas.v6n1p129.

Agelet, A., Bonet, M.À. and Vallés, J., 2000. Homegardens and their role as a main source of medicinal plants in mountain regions of Catalonia (Iberian peninsula). *Economic Botany*, 54, 295–309. https://doi.org/10.1007/BF02864783.

Aguilar-Støen, M., Moe, S.R., Camargo-Ricalde, S.L., 2009. Home gardens sustain crop diversity and improve farm resilience in Candelaria Loxicha, Oaxaca, Mexico. *Human Ecology*, 37, 55–77. https://doi.org/10.1007/s10745-008-9197-y.

Ahmad, M., Nawab, K., Zaib, U. and Khan, I.A., 2007. Role of women in vegetable production: A case study of four selected villages of district Abbottabad. *Sarhad Journal of Agriculture*, 23(8), pp. 1174–1180.

Alaimo, K., Packnett, E., Miles, R.A. and Kruger, D.J., 2008. Fruit and vegetable intake among urban community gardeners. *Journal of Nutrition Education and Behavior*, 40(2), pp. 94–101. https://doi.org/10.1016/j.jneb.2006.12.003.

Ali, A.M.S., 2005. Homegardens in small holder Farming systems: Examples from Bangladesh. *Human Ecology*, 33, pp. 245–270. https://doi.org/10.1007/s10745-005-2434-8.

Alon-Mozes, T., 2007. Rooted in the home garden and in the nation's landscape: Women and the emerging Hebrew garden in Palestine. *Landscape Research*, 32, pp. 311–331. https://doi.org/10.1080/01426390701318197.

Alonso, E.B., Cockx, L. and Swinnen, J. 2018. Culture and food security. *Global Food Security*, 17, pp. 113–127. https://doi.org/10.1016/j.gfs.2018.02.002.

Augustino, S. and Gillah, P.R., 2005. Medicinal plants in urban districts of Tanzania: Plants, gender roles and sustainable use. *The International Forestry Review*, 7, pp. 44–58.

Bain, J.H., 1993. Mexican rural women's knowledge of the environment. *Mexican Studies/Estudios Mexicanos*, 9, pp. 259–274. https://doi.org/10.2307/1051879.

Bajwa, B.E., Aslam, M.N. and Malik, A.H., 2015. Food security and socio-economic conditions of women involved in kitchen gardening in Muzaffargarh, Punjab, Pakistan. *Journal of Enviornmental and Agricultural Studies*, 4.

Ban, N. and Coomes, O.T., 2004. Home gardens in Amazonian Peru: Diversity and exchange of planting material. *Geographical Review*, 94, pp. 348–367.

Bargali, K., 2015. Comparative participation of rural women in agroforestry home gardens in Kumaun Himalaya, Uttarakhand, India. *Asian Journal of Agricultural Extension, Economics & Sociology*, 6, pp. 16–22. https://doi.org/10.9734/AJAEES/2015/16115.

Bhatti, M. and Church, A., 2004. Home, the culture of nature and meanings of gardens in late modernity. *Housing Studies*, 19, pp. 37–51. https://doi.org/10.1080/026730304200 0152168.

Blaine, T.W., Grewal, P.S., Dawes, A. and Snider, D., 2010. Profiling community gardeners. *Journal of Extension*, 48(6), no. 6FEA6.

Buckingham, S., 2005. Women (re)construct the plot: The regen(d)eration of urban food growing. *Area*, 37, pp. 171–179. https://doi.org/10.1111/j.1475-4762.2005.00619.x.

Bushamuka, V.N., Pee, S. de, Talukder, A., Kiess, L., Panagides, D., Taher, A. and Bloem, M., 2005. Impact of a homestead gardening program on household food security and empowerment of women in Bangladesh. *Food and Nutrition Bulletin*, 26, pp. 17–25.

Chambers, K.J. and Momsen, J.H., 2007. From the kitchen and the field: Gender and maize diversity in the Bajío region of Mexico. *Singapore Journal of Tropical Geography*, 28, pp. 39–56. https://doi.org/10.1111/j.1467-9493.2006.00275.x.

Chauhan, N.M., 2012. Impact and constraints faced by tribal farm women in kitchen gardening. *Rajasthan Journal of Extension Education*, 20, pp. 87–91. http://www.rseeudaipur.org/wp-content/uploads/2013/02/223.pdf.

Coughlan, R. and Hermes, S.E., 2016. The palliative role of green space for Somali Bantu women refugees in displacement and resettlement. *Journal of Immigrant & Refugee Studies*, 14, pp. 141–155. https://doi.org/10.1080/15562948.2015.1039157.

Faber, M., Phungula, M.A., Venter, S.L., Dhansay, M.A. and Benadé, A.S., 2002. Home gardens focusing on the production of yellow and dark-green leafy vegetables increase the serum retinol concentrations of 2–5-y-old children in South Africa. *American Journal of Clinical Nutrition*, 76, pp. 1048–1054. https://doi.org/10.1093/ajcn/76.5.1048.

Finerman, R. and Sackett, R., 2003. Using home gardens to decipher health and healing in the Andes. *Medical Anthropology Quarterly*, 17, pp. 459–482. https://doi.org/10.1525/maq.2003.17.4.459.

Food and Agriculture Organization of the United Nations, 2001. *The state of food insecurity in the world 2001*. Rome, Italy: Food and Agriculture Organization of the United Nations.

Food and Agriculture Organization of the United Nations, 2011. *The state of food and agriculture: Women in agriculture: Closing the gender gap for development*. Rome, Italy: Food and Agriculture Organization of the United Nations.

Galhena, D.H., Freed, R. and Maredia, K.M., 2013. Home gardens: A promising approach to enhance household food security and wellbeing. *Agriculture & Food Security*, 2, p. 8. https://doi.org/10.1186/2048-7010-2-8.

Girard, A.W., Self, J.L., McAuliffe, C. and Olude, O., 2012. The effects of household food production strategies on the health and nutrition outcomes of women and young children: A systematic review. *Paediatric and Perinatal Epidemiology*, 26, pp. 205–222. https://doi.org/10.1111/j.1365-3016.2012.01282.x.

Gowdy-Wygant, C., 2013. *Cultivating victory: The women's land army and the victory garden movement*. University of Pittsburgh Press. https://doi.org/10.2307/j.ctt9qh5qk.

Gray, L., Guzman, P., Glowa, K.M. and Drevno, A.G., 2014. Can home gardens scale up into movements for social change? The role of home gardens in providing food security and community change in San Jose, California. *Local Environment*, 19, pp. 187–203. https://doi.org/10.1080/13549839.2013.792048.

Guitart, D., Pickering, C. and Byrne, J., 2012. Past results and future directions in urban community gardens research. *Urban Forestry and Urban Greening*, 11(4), pp. 364–373. http://dx.doi.org/10.1016/j.ufug.2012.06.007.

Harris, N., Minniss, F.R. and Somerset, S., 2014. Refugees connecting with a new country through community food gardening. *International Journal of Environmental Research and Public Health*, 11, pp. 9202–9216. https://doi.org/10.3390/ijerph110909202.

Horenstein, N.R., 1989. Women and food secruity in Kenya. *Working Paper No. WPS 232.* Women in Development. The World Bank.

Howard, P.L., 2003. *The major importance of "minor" resources: Women and plant biodiversity* (No. 112), GateKeeper Series. London, UK: International Institute for Environment and Development.

Howard, P.L., 2006. Gender and social dynamics in swidden and homegardens in Latin America. In B.M. Kumar and P.K.R. Nair, eds. *Tropical homegardens: A time-tested example of sustainable agroforestry, advances in agroforestry*. Dordrecht, Netherlands: Springer, pp. 159–182. https://doi.org/10.1007/978-1-4020-4948-4_10.

Howard-Borjas, P. and Cuijpers, W., 2009. Gender relations in local plant genetic resource management and conservation. In H.W. Doelle, J.S. Rokem and M. Berovic, eds. *Biotechnology, volume XIV: Fundamentals in biotechnology*. Oxford, UK: EOLSS Publications, pp. 20–58.

Keys, E., 1999. Kaqchikel gardens: Women, children and multiple roles of gardens among the maya of Highland Guatemala: Yearbook. *Conference of Latin Americanist Geographers*, 25, pp. 89–100.

Kumar, B.M. and Nair, P.K.R., 2004. The enigma of tropical homegardens. In P.K.R. Nair, M.R. Rao and L.E. Buck, eds. *New vistas in agroforestry: A compendium for 1st world congress of agroforestry, 2004, advances in agroforestry*. Dordrecht, Netherlands: Springer, pp. 135–152. https://doi.org/10.1007/978-94-017-2424-1_10.

Marambe, B. et al., 2018. Climate variability and adaptation of Homegardens in South Asia: Case studies from Sri Lanka, Bangladesh and India, Sri Lanka. *Sri Lanka Journal of Food and Agriculture*, 4(2), pp. 7–27. DOI: 10.4038/sljfa.v4i2.61.

Márquez, A.R.C. and Schwartz, N.B., 2008. Traditional home gardens of Petén, Guatemala: Resource management, food security and conservation. *etbi*, 28, pp. 305–317. https://doi.org/10.2993/0278-0771-28.2.305.

Mattalia, G., Calvo, A. and Migliorini, P., 2018. Alpine home gardens in the Western Italian Alps: The role of gender on the local agro-biodiversity and its management. *Biodiversity*, 19, pp. 179–187. https://doi.org/10.1080/14888386.2018.1504692.

Mattsson, E., Ostwald, M. and Nissanka, S., 2018. What is good about Sri Lankan homegardens with regards to food security? A synthesis of the current scientific knowledge of a multifunctional land-use system. *Agroforestry Systems*, 92(6), pp. 1469–1484. https://doi.org.proxy1.cl.msu.edu/10.1007/s10457-017-0093-6.

Mazumdar, S. and Mazumdar, S., 2012. Immigrant home gardens: Places of religion, culture, ecology and family. *Landscape and Urban Planning*, 105, pp. 258–265. https://doi.org/10.1016/j.landurbplan.2011.12.020.

Méndez, V.E., Lok, R. and Somarriba, E., 2001. Interdisciplinary analysis of homegardens in Nicaragua: Micro-zonation, plant use and socioeconomic importance. *Agroforestry Systems*, 51, pp. 85–96. https://doi.org/10.1023/A:1010622430223.

Murrieta, R.S.S. and WinklerPrins, A.M.G.A., 2003. Flowers of water: Homegardens and gender roles in a Riverine Caboclo community in the lower amazon, Brazil. *Culture & Agriculture*, 25, pp. 35–47. https://doi.org/10.1525/cag.2003.25.1.35.

Ngome, I. and Foeken, D., 2012. "My garden is a great help": Gender and urban gardening in Buea, Cameroon. *GeoJournal*, 77, pp. 103–118. https://doi.org/10.1007/s10708-010-9389-z.

Oakley, E. and Momsen, J.H., 2005. Gender and agrobiodiversity: A case study from Bangladesh. *The Geographical Journal*, 171, pp. 195–208. https://doi.org/10.1111/j.1475-4959.2005.00160.x.

Olum, S., Ongeng, D., Tumuhimbise, G.A., Hennessy, M.J., Okello-Uma, I. and Taylor, D., 2018. Understanding intra-community disparity in food and nutrition security in a generally food insecure part of eastern Africa. *African Journal of Food, Agriculture, Nutrition and Development*, 18(2), pp. 13317–13337. http://article.foodnutritionresearch.com/pdf/jfnr-5-6-10.pdf.

Parry, D.C., Glover, T.D. and Shinew, K.J., 2005. "Mary, mary quite contrary, how does your garden grow?" Examining gender roles and relations in community gardens. *Leisure Studies*, 24, pp. 177–192. https://doi.org/10.1080/0261436052000308820.

Patalagsa, M.A., Schreinemachers, P., Begum, S. and Begum, S., 2015. Sowing seeds of empowerment: Effect of women's home garden training in Bangladesh. *Agriculture & Food Security*, 4, p. 24. https://doi.org/10.1186/s40066-015-0044-2.

Rammohan, A., Pritchard, B. and Dibley, M., 2019. Home gardens as a predictor of enhanced dietary diversity and food security in rural Myanmar. *BMC Public Health*, 19(1145), p. 13. https://doi-org.proxy1.cl.msu.edu/10.1186/s12889-019-7440-7.

Raney, T. et al., 2011. The role of women in agriculture. *ESA Working Paper No. 11–20*. Rome, Italy: The Food and Agriculture Organization of the United Nations. Available at: www.fao.org/3/am307e/am307e00.pdf.

Reyes-García, V., Vila, S., Aceituno-Mata, L., Calvet-Mir, L., Garnatje, T., Jesch, A., Lastra, J.J., Parada, M., Rigat, M., Vallès, J., Pardo-de-Santayana, M., 2010. Gendered homegardens: A study in three mountain areas of the Iberian Peninsula. *Economic Botany*, 64, pp. 235–247.

Rowe, W.C., 2009. "Kitchen gardens" in Tajikistan: The economic and cultural importance of small-scale private property in a post-soviet society. *Human Ecology*, 37, pp. 691–703.

Rubaihayo, E.B., 2002. Uganda – The contribution of indigenous vegetables to household food security. *Indigenous Knowledge (IK) Notes*, No. 44. World Bank, Washington, DC. https://openknowledge-worldbank-org.proxy1.cl.msu.edu/handle/10986/10794.

Schreinemachers, P. et al., 2015. The effect of women's home gardens on vegetable production and consumption in Bangladesh. *Food Security*, 7(1), pp. 97–107. DOI: 10.1007/s12571-014-0408-7.

Shillington, L., 2008. Being(s) in relation at home: Socio-natures of patio "gardens" in Managua, Nicaragua. *Social & Cultural Geography*, 9, pp. 755–776. https://doi.org/10.1080/14649360802382560.

Strunk, C. and Richardson, M., 2019. Cultivating belonging: Refugees, urban gardens and placemaking in the Midwest, U.S.A. *Social & Cultural Geography*, 20, pp. 826–848. https://doi.org/10.1080/14649365.2017.1386323.

Torri, M.C., 2012. Mainstreaming local health through herbal gardens in India: A tool to enhance women active agency and primary health care? *Environment, Development and Sustainability: Dordrecht*, 14, pp. 389–406. http://dx.doi.org.proxy2.cl.msu.edu/10.1007/s10668-011-9331-7.

Trinh, L.N., Watson, J.W., Hue, N.N., De, N.N., Minh, N.V., Chu, P., Sthapit, B.R. and Eyzaguirre, P.B., 2003. Agrobiodiversity conservation and development in Vietnamese home gardens. *Agriculture, Ecosystems & Environment*, 97, pp. 317–344. https://doi.org/10.1016/S0167-8809(02)00228-1

van den Bold, M., Quisumbing, A.R. and Gillespie, S., 2013. Women's empowerment and nutrition: An evidence review. *International Food Policy Research Institute*. Discussion Paper 01294, pp. 1–80.

Vazquez-Garcia, V., 2008. Gender, ethnicity and economic status in plant management: Uncultivated edible plants among the Nahuas and Populucas of Veracruz, Mexico. *Agriculture and Human Values: Dordrecht*, 25, p. 65. http://dx.doi.org.proxy2.cl.msu.edu/10.1007/s10460-007-9093-x.

Vibert, E., 2016. Gender, resilience and resistance: South Africa's Hleketani Community Garden. *Journal of Contemporary African Studies*, 24, pp. 252–267. https://doi.org.proxy1.cl.msu.edu/10.1080/02589001.2016.1202508.

Waliczek, T.M., Mattson, R.H. and Zajicek, J.M., 1996. Benefits of community gardening on quality of life issues. *Journal of Environmental Horticulture*, 14(4), pp. 204–209. www.hrijournal.org/doi/pdf/10.24266/0738-2898-14.4.204.

Watson, J.W. and Eyzaguirre, P.B. eds., 2001. Home gardens and in situ conservation of plant genetic resources in farming systems. In *Second international home gardens workshop*. Witzenhausen: Federal Republic of Germany.

Wegren, S.K., Nikulin, A., Trotsuk, I. and Golovina, S., 2017. Gender inequality in Russia's rural informal economy. *Communist and Post: Communist Studies*, 50, pp. 87–98. https://doi.org/10.1016/j.postcomstud.2017.05.007.

Wilson-Moore, M., 1989. Women's work in homestead gardens: Subsistence, patriarchy and status in Northwest Bangladesh. *Urban Anthropology and Studies of Cultural Systems and World Economic Development*, 18, pp. 281–297.

World Bank, 2019. *World Bank open data*. [Online] Available at: https://data.worldbank.org/ [Accessed 14 November 2019].

Yasmin, T., Khattak, R. and Ngah, I., 2013. Facilitating earthquake: Affected rural women communities toward sustainable livelihoods and agriculture. *Agroecology and Sustainable Food Systems*, 37, pp. 592–613. https://doi.org/10.1080/21683565.2012.762637.

Yasmin, T., Khattak, R. and Ngah, I., 2014. Eco-friendly kitchen gardening by Pakistani rural women developed through a farmer field school participatory approach. *Biological Agriculture & Horticulture*, 30, pp. 32–41. https://doi.org/10.1080/01448765.2013.845112.

5 Home gardens for better health and nutrition in Mozambique

Adrienne Attorp, Amir Kassam and Peter Dorward

Introduction

Overcoming global hunger and food insecurity remains one of mankind's greatest challenges. In sub-Saharan Africa, one in four persons is chronically hungry (FAO, IFAD and WFP, 2015). Further, "hidden hunger" or micronutrient deficiency, an often overlooked component of hunger, continues to affect over two billion people worldwide (von Grebmer et al., 2014). In Cabo Delgado, Mozambique, where more than 50% of the population lives below the poverty line, this poses a significant health challenge (International Bank for Reconstruction and Development, 2009).

While in recent decades there has been much focus on the issue of improving food security in Mozambique and other developing countries by improving production or supply of food and access to food through employment, the importance of addressing the nutritional quality of diets, a component of food security, has often not been fully recognized (von Grebmer et al., 2014; Keatinge et al., 2011; Iannotti et al., 2009). Even when households may have access to sufficient food to meet their caloric requirements, their diets are often of poor quality – high in staple starches and not very diverse. As a result, many people fail to consume adequate vitamins and minerals, leaving them at high risk of micronutrient deficiencies and vulnerable to ill health and diseases (von Grebmer et al., 2014; Keatinge et al., 2011; Iannotti et al., 2009).

Compounding this problem, in much of sub-Saharan Africa, rural households often suffer a "hungry gap" of two to four months toward the end of the dry season and beginning of the wet season. During this time, many farm families and rural people, including women and children, can suffer from a diet that is inadequate not only in terms of micronutrients but also calories and protein. For those living in poverty, repeated exposure to such seasonal stresses can undermine their ability to cope in the long-term (Sullivan, 2013). Long-term hidden hunger and recurrent hungry gaps can have irreversible health effects and negative socioeconomic consequences for individuals and families, which impact their well-being and development (von Grebmer et al., 2014).

Further, challenges such as rising food prices, climate change, political instability and competition for land and natural resources increasingly put pressure on individuals and families as they try to earn a living and be "food secure," defined by the World Health Organization (2015) as having access to "sufficient safe, nutritious food to maintain a healthy and active life" (FAO, IFAD and WFP, 2014; von Grebmer et al., 2014; Keatinge et al., 2011).

One approach that has proven successful in increasing micronutrient consumption, diversifying diets and improving food availability is the implementation of home gardening programs (von Grebmer et al., 2014; DFID, 2014; Masset et al., 2012; Cabalda et al., 2011; Ecker et al., 2010; Iannotti et al., 2009; Yiridoe and Anchirinah, 2005; Eyzaguirre and Linares, 2004; Krishna, 2004; Sthapit et al., 2004; Faber et al., 2002a, 2002b; Bloem et al., 1996). In Cabo Delgado, the Aga Khan Foundation (AKF), an international non-governmental organization (NGO), has been implementing such a program since 2009. Using this intervention as a case study, field research was conducted to investigate the feasibility of using home gardens as a tool to address the high rates of food insecurity and malnutrition in Cabo Delgado Province.

This chapter outlines many of the current food security and malnutrition issues faced by the people of Cabo Delgado Province, as well as the rural development initiative implemented by the AKF in the region. Findings of the research are then discussed in the context of the broad malnutrition and food insecurity issues that face many populations around the world, as well as other research that has been conducted on home gardening practices and interventions.

Nutrition

Good nutrition is a cornerstone of health and well-being: fundamental to physical and cognitive development and critical to economic productivity (FAO, IFAD and WFP, 2014; von Grebmer et al., 2014). The high rate of malnutrition experienced in developing countries critically affects the development progress (FAO, IFAD and WFP, 2014; von Grebmer et al., 2014). There are also a variety of health and social consequences that can be directly linked to poor nutrition across the lifespan, particularly in childhood (FAO, IFAD and WFP, 2014; Harris, 2011; Keating et al., 2011; Iannotti et al., 2009; Black et al., 2008).

Micronutrient deficiencies increase the risk of mortality from diarrhea, pneumonia, malaria and other illnesses, as well as the risk of developing a host of other diseases and developmental issues (von Grebmer et al., 2014; Keating et al., 2011; West, 2003). This is an issue in the developing world in particular, where poor populations tend to rely on nutrient-poor, starchy staples such as rice, maize and cassava to make up the bulk of their diet (AKF, 2013; Ecker et al., 2010). Such deficiencies (vitamin A and iron being the most prevalent) are responsible for a large proportion of infections, as well as poor physical and mental development and excess mortality (FAO, 2014; Hotz et al., 2012; Keatinge et al., 2011; Harris, 2011; Ecker et al., 2010; Iannotti et al., 2009).

Increasing dietary diversity is one of the most effective ways to sustainably prevent hidden hunger (Thompson and Amoroso, 2010). In particular, the fruit and vegetable food group can significantly contribute to the mitigation of common micronutrient deficiencies and play an important role in supporting overall good health. Indeed, Keating et al. (2011) state that globally there is a highly significant association between vegetable availability per person and the mortality rate of children under five.

Implementation of interventions aiming to increase fruit and vegetable consumption as a way of mitigating micronutrient deficiencies remains relevant. Home gardening interventions have been shown to be one of the most successful of these.

Home gardening – what is a home garden?

The practice of home gardening is widespread in the developing world and takes place in a diverse range of countries and societies. A home garden (HG) can be defined in various ways, but is typically assumed to involve the management of a variety of different plant and (in many cases) animal species and is generally located close to permanent or semi-permanent dwellings (Eyzaguirre and Linares, 2004; Krishna, 2004; Sthapit et al., 2004). Sthapit et al. (2004) define a home garden as "a micro-environment composed of a multi-species (annual to perennial, root crops to climbers, etc.), multi-storied and multi-purpose garden situated close to the homestead." It serves to supply and supplement subsistence requirements and may also generate secondary income (Keatinge et al., 2011; Yiridoe and Anchirinah, 2005; Eyzaguirre and Linares, 2004). However, its products are primarily intended for household consumption (Sthapit et al., 2004). Plants grown in home gardens have multiple purposes, producing outputs such as annual and perennial food crops, medicines, fuel, construction material and animal fodder. These may either be used by the household or sold for income (Cabalda et al., 2011; Keatinge et al., 2011; Yiridoe and Anchirinah, 2005; Eyzaguirre and Linares, 2004; Sthapit et al., 2004).

Home gardens are of no fixed size and can be found in both urban and rural settings. They may be either static or moving; shifting cultivation of field crops is a common practice in many tropical regions (Eyzaguirre and Linares, 2004; Sthapit et al., 2004), but home gardens are most commonly continuously cultivated and are maintained by members of the household (Sthapit et al., 2004). Note that other types of gardens, such as community food growing projects or school gardens have been found to have many similar outcomes for health and nutrition as do those classified strictly as home gardens (The Edible School Yard Project, 2015; Soil for Life, 2013; Rauzon et al., 2010; Nutrition and Consumer Protection Division of FAO, 2010).

Home gardens and nutrition

Various studies from around the world highlight the importance of the contributions home gardens make to the nutrition and livelihoods of individuals and communities (DFID, 2014; Selepe and Hendriks, 2014; Masset et al.,

2012; Cabalda et al., 2011; Ecker et al., 2010; Keatinge et al., 2011; Talukder et al., 2010; Iannotti et al., 2009; Laurie and Faber, 2008; Musotsi et al., 2008; Yiridoe and Anchirinah, 2005; Eyzaguirre and Linares, 2004; Krishna, 2004; Sthapit et al., 2004; Faber et al., 2002a, 2002b; Bloem et al., 1996). The implementation of home gardening interventions in regions vulnerable to malnutrition has proven to be successful in both increasing and diversifying fruit and vegetable consumption in target populations. In light of their potential to significantly improve nutrition outcomes, such programs have become an integral component of a variety of nutrition intervention programs being implemented by development agencies around the world, including Concern, Helen Keller International (HKI) and many others (Concern Worldwide, 2009; Iannotti et al., 2009). Coupling such interventions with nutrition education programs has proven to be especially effective (AKF, 2013; Keating et al., 2011; Iannotti et al., 2009; Faber et al., 2002a, 2002b).

In fact, in a review of agricultural interventions aimed at improving nutrition outcomes, Berti et al. (2004) conclude that home gardening projects had a higher success rate with respect to increasing dietary diversity and micronutrient intake than that of any other type of intervention considered. Moreover, home gardens are easily transferable and scalable, making them a cost-effective and adaptable intervention (Keatinge et al., 2011).

The diverse array of foods produced in home gardens is a vital source of essential nutrients and can contribute significantly to food security (Keatinge et al., 2011; Iannotti et al., 2009; Musotsi et al., 2008; Yiridoe and Anchirinah, 2005; Eyzaguirre and Linares, 2004; Krishna, 2004; Sthapit et al., 2004; Faber et al., 2002a, 2002b). Keatinge et al. (2011) and Sthapit et al. (2004) highlight their function as a safety net for households when food is scarce. Krishna (2004) notes also that consumption of foods grown in home gardens is facilitated because of the fact that they are typically culturally preferred.

Further to this, Keatinge et al. (2011) and Eyzaguirre and Linares (2004) note that the resources home gardens provide can act as an important buffer against fluctuations in the cash economy, and the year-round availability of food, fuel, feed and construction materials contributes significantly to economic stability. HKI has found that villages that adopt homestead gardening, along with the raising of small livestock, are significantly better off in terms of nutrition and prosperity as compared to neighboring control villages (Talukder et al., 2010).

Iannotti et al. (2009) describe the contribution home gardens make to nutrition in terms of food security: increased food availability, measured by changes in the amounts and varieties of food products produced; increased food access, measured by household level consumption as well as expenditures on micronutrient-rich foods; and increased food utilization, measured by individual intakes of micronutrient-rich foods as well as physical indicators of micronutrient status (e.g., anthropometry).

In light of the fact that a diverse diet acts as an important buffer against a variety of diseases, it can be seen that the contributions home gardens make to diet diversity is of great importance, particularly for low-income households

(Keatinge et al., 2011; Faber and Laurie, 2010). For example, in eastern Africa, Ecker et al. (2010) found that vegetables were the primary source of vitamin A for almost all households studied in Rwanda, Tanzania and Uganda. Households with home gardens consumed more fruits and vegetables, and it was found that home gardens had a significant positive effect on vitamin A intake in all samples.

Research from South Africa conducted by Faber et al. (2002a, 2002b) conducted in rural South Africa found similar results. One year after a home gardening program was implemented as part of a broader nutrition intervention initiative – using demonstration gardens as training centers for nutrition and gardening and developing community-based plant nurseries which produce and distribute orange sweet potato cuttings (rich in vitamin A) and seedlings to community members – children from households with a home garden demonstrated significantly higher vitamin A intake than children from homes without gardens. Subsequent research by Faber and Laurie (2010) further supports the positive outcomes achieved in this intervention, showing that among program participants there were improvements in maternal knowledge of vitamin A nutrition, dietary intake of vegetables rich in vitamin A, caregiver-reported child morbidity and children's vitamin A status.

Role of education

It is essential to highlight the importance of coupling nutrition education with the increases in production achieved via such interventions (AKF, 2013; Hotz et al., 2012; Faber and Laurie, 2010; Iannotti et al., 2009; Faber et al., 2002a, 2002b). Iannotti et al. (2009, p. 5) state: "nutrition education is necessary to translate food production into improved dietary intakes." This should include both specific nutritional objectives and nutrition education activities.

Webb and Rogers (2003, p. 5) discuss nutrition knowledge as a key component of what they refer to as 'nutrition capital' and emphasize the important role it plays in contributing to improving food utilization. They indicate that nutrition knowledge "incorporates issues of food safety and quality, sufficiency of intake at the individual level and the conversion efficiency of food by the body that result in sound nutritional status and growth." Cabalda et al. (2011) provide further support for this idea, stating that nutrition education may be an important means of changing eating and feeding practices among certain populations.

Home gardens remain an important source of nutrition and livelihood for many in the developing world and can be an effective component of international development projects, as evidence shows. For this reason, AKF has included a home gardening initiative as part of its greater development program being implemented in northern Mozambique. It is this initiative that forms the basis for research conducted (Attorp, 2011) and reported here. Following is a discussion of AKF's program and the region in which it is being implemented.

Case study from Cabo Delgado, Mozambique

With an area of 800,000 km² and a population of only 27.2 million people, Mozambique is one of the largest and least densely populated countries in southern Africa (World Bank Group, 2015). Seventy percent of its population lives rurally (UN, 2010).

In 2014, Mozambique ranked 178 out of 187 countries on the United Nations Human Development Index (HDI; UN, 2014). It is estimated that more than 54% of the population lives in poverty, with a per capita gross domestic product (GDP) of only USD 602.10 in 2014 (World Bank Group, 2015). It is in rural areas where poverty is most acute.

Cabo Delgado is the northernmost of Mozambique's ten provinces, bordering southern Tanzania, along the coast. It has among the highest poverty rates in the country, with 63% of the province's 1.6 million people living below the poverty line (AKF, 2013). It is in this province that the Aga Khan Foundation Mozambique (AKF(M)), in conjunction with the Canadian International Development Agency (CIDA), is implementing the Enhancing Food Security and Increasing Incomes initiative (FSI), a six-year project which aims to improve the overall quality of life of Cabo Delgado's population (AKFC, 2011). The project is being implemented in seven districts, with an overall goal of improving food security and livelihood opportunities for up to 35,000 households (AKFC, 2011).

Cabo Delgado: the population

According to an extensive survey undertaken by the AKF in Cabo Delgado in 2009 (AKF, 2009), the population of Cabo Delgado Province is approximately 1.61 million people (at the time of the survey), with 1.15 million (71.4%) of those rural. Mean household size was estimated at 5.6 in urban areas and 5.0 in rural areas.

Much of the province's population lives in isolated villages which have poor access to year-round passable roads and, by extension, poor access to markets, education, health services, and power and communication infrastructure (AKFC, 2011). According to AKF(M)'s 2009 survey, in most areas, including those related to mortality, nutritional status, education and child and maternal health, Cabo Delgado's indicators are worse than the national average.

For most households in the region, the main source of food and income is the *machamba* – rural smallholdings where a variety of staple and horticultural crops are grown (AKFC, 2011). Productivity of these farms is severely constrained by a variety of interconnected factors, including poor infrastructure, weak input and output market systems, unaffordable credit and lack of access to information about improving marketing and production practices. Further constraints associated with illiteracy, poverty and geographical isolation compound these issues (AKFC, 2011).

Food insecurity has historically been rife in many parts of Cabo Delgado as a result of cyclical droughts and flooding. On average, 34% of households in the region are food insecure and face chronic hunger (AKFC, 2011). A 2008 assessment undertaken by the Aga Khan Development Network (AKDN) found that only 3.1%–6.5% of surveyed households were considered food secure during the region's lean season (January and February; AKF, 2013).

High levels of food insecurity are reflected in the fact that of children younger than five years old from rural areas, 52.7% are stunted (Demographic and Health Surveys Program, 2011). These rates are some of the highest in the country, and it is not a coincidence that Cabo Delgado also has the highest child mortality rates in Mozambique (AKF, 2009). According to AKF, these higher-than-average rates of mortality are due to the fact that the majority of the province's population has little access to basic health services (AKF, 2009).

With respect to micronutrient deficiencies, the picture is equally dire. While no data specific to Cabo Delgado is available on the matter, according to United Nations Children's Fund (2011) 75% of children in Mozambique are iron deficient and 69% are vitamin A deficient. For mothers, the figures are 48% and 11%, respectively.

AKF food security and increasing incomes initiative: home gardening component

In light of the malnutrition issues faced by Cabo Delgado's rural population, AKF is currently implementing a Food Security and Increasing Incomes Initiative (FSI) in the region. The project comprises a variety of development initiatives, as described in the project implementation plan (AKFC, 2011).

The food security component of the project has a broad, intermediate outcome of increasing adoption of improved nutritional practices among the population. This intermediate outcome has two immediate outcomes, one of which is to improve awareness of nutrition practices among both women and men. The establishment of home gardens is one of three planned outputs implemented in support of this immediate outcome (AKFC, 2011). Research for this study occurred in conjunction with the home gardening initiative.

In order to support the uptake and proliferation of home gardening (HG) practices, regional and local agriculture facilitators working for AKFM have provided training and extension services primarily to existing community-based nutrition groups. These groups, which are supported by the health component of the FSI, have been established as part of the greater FSI initiative. However, home gardening practices are also being promoted to a lesser extent through farmer field schools, agriculture associations and village development organizations (AKFC, 2011).

Nutrition groups consist of approximately ten member households, each of which has been identified by village leaders as being a good candidate for participation in the program. There is one group per village (AKFC, 2011). The groups are provided with both education on good nutrition practices and with nutrition message training, so that this nutrition and health information can be disseminated to other households within their communities. It is intended that, through this model, home gardening practices will be promoted and supported within each village as well (AKFC, 2011).

Demonstration gardens have been established in each district in order to facilitate this training. At the time of writing, nine villages in three of the seven districts in which the FSI is being implemented had been targeted. Here there were 82 gardens recorded, with 45 of these being demonstration plots.

AKF(M)'s home gardening project was new for the 2011 growing season; gardens observed as part of this research had not yet been in production for a full season. The program's target for the 2011 season was to have 150 households with home gardens by the end of the season. The overall target for the program is to have 1,500 households with home gardens in five of the seven districts.

Simple, easy to maintain gardens are being promoted (approximately 4 m^2 in size) in which high-nutrient, drought-resistant vegetables are to be grown (AKFC, 2011; based on discussions with AKF(M) agriculture facilitators and supervisors, 2011). It is hoped that, in addition to improving food security by facilitating access to an increased number of vegetables, such home gardens will play a key role in improving the nutritional status of people in these communities – women and children in particular (AKFC, 2011).

In similar research, a minimum garden size of 6 m^2 or greater is promoted (e.g., Faber and Laurie, 2010; Bhattacharjee et al., 2006) and as discussed earlier, Diana et al. (2014) indicate that in their research small garden size may have contributed to poor nutrition outcomes. However, it remains to be seen whether small garden size may negatively impact nutrition outcomes in this study.

According to AKF(M)'s Project Implementation Plan (2011), with respect to crops that are to be promoted for the home gardening project, priority will be given to [orange] sweet potato for its vitamin A content, pumpkin for its long storage potential, moringa for its high vitamin content and drought resistance, and sesame for its high iron content (AKFC, 2011). Consumption of the leaves of these vegetables will also be promoted, as they too are rich in vitamins (AKFC, 2011).

Research for this study was conducted in conjunction with AKF(M)'s home gardening project. The primary goal of research was to determine the feasibility of using these home gardens as a tool to address the high rates of food insecurity and malnutrition among women and children in Cabo Delgado Province. In particular, issues pertaining specifically to vegetable consumption were investigated, given the high potential that vegetable production and consumption has to mitigate micronutrient deficiencies.

Research was conducted among Cabo Delgado's rural population. Over the course of three weeks in June and July of 2011, five villages in three different districts were visited, where a combination of focus groups, participatory research activities (a social mapping exercise and a nutrition calendar) and personal interviews were undertaken. A detailed description of methodology used is given in Attorp (2011).

Participants were classified as having a home garden if they had a home garden at the time of the interview or had started one at some point during that season. Individuals were considered not to have a home garden only if they had not had one at any point during that season.

Focus groups and personal interviews for individuals with home gardens centered around the perceptions that participants held about home gardening, as well as their experiences with the practice. This included a discussion of the benefits they had experienced since adopting the practice, as well as the barriers they faced in starting and maintaining their home gardens. Further, if any harvesting of produce had occurred, discussion of what was done with the produce took place.

Discussion with individuals without home gardens again focused on perceptions about the practice. In particular, reasons for their non-adoption were investigated. Further, individuals were asked hypothetical questions about the challenges and benefits they thought they would experience should they start a home garden.

For both groups, issues pertaining to nutrition were also discussed. Topics included nutrition habits (i.e., what was eaten), consumption practices (i.e., who eats what) and perceptions about healthful foods.

Results and discussion

Nutrition

It has been established that there is a significant malnutrition problem in Cabo Delgado Province, particularly among women and children (AKF, 2013, 2011, 2009). Given that malnutrition can have a number of negative consequences for health and social outcomes across the lifespan, the need to address this issue is clear (FAO, 2014; Keatinge et al., 2011).

As various authors have discussed, the problem of malnutrition is a broad and multifaceted one, with a number of different contributing causes (FAO, IFAD and WFP, 2014; von Grebmer et al., 2014; Sen, 1981). Many of these are common among those facing malnutrition. However, without a more nuanced picture of what is causing malnutrition in a given area, it is difficult to address the problem in an effective manner.

This study chose to investigate issues pertaining specifically to vegetable consumption in Cabo Delgado in an attempt to better understand some of the causes of malnutrition problems in this region. Through discussion with individuals in various districts in the region, certain common issues began to emerge.

Vegetable consumption in Cabo Delgado Province

Food items most commonly consumed by study participants throughout the year were identified through use of a participatory research activity: creation of a nutritional calendar. In line with evidence from AKF's baseline study (2011) and from similar studies (von Grebmer et al., 2014; Keatinge et al., 2011; Faber and Laurie, 2010), it was found the majority of study participants relied heavily on starches as the staples of their diet year-round: cassava, bread, rice, maize and millet. Dietary diversity varied from location to location, however in all cases fairly significant seasonal variation was present, with vegetables, animal protein and legumes consumed on a seasonal basis only, if at all.

In many cases, vegetables, aside from some leafy greens (*matapa* – moringa, sweet potato or bean leaves), were consumed very rarely. In all instances, January and February were highlighted as the months when food was least varied – an outcome which was not unexpected given that those are known as the "hungry months" in the region (AKF, 2013). These results are in line with previous research that indicates vegetable consumption among many populations worldwide is not meeting recommended guidelines (Keatinge et al., 2011; Ecker et al., 2010).

However, despite the fact that data collected in the nutrition calendar indicated vegetables were not a predominant fixture in the diets of villagers in Cabo Delgado, the consensus among the majority of study participants was that matapa and other vegetables are good for health. In most cases, respondents indicated that this was because vegetables are high in vitamins.

Conversely, in the majority of responses, most of the staple starches that form the base of most people's diet in the region (cassava, rice, wheat) were deemed to have little nutritional value. Only maize was identified in some cases to be healthful. According to one focus group member: "The cassava flour when we eat it makes our stomachs very heavy [full], but it doesn't give us any vitamins."

Such results are positive and bode well for the intervention as it attempts to increase access to and consumption of vegetables in the region. However, it should be noted there was some confusion on the matter of what is healthful or not. For example, not everyone was certain whether or not vegetables needed to be cooked in oil in order to be healthy, or whether other foods (e.g., fish) were healthy at all. This is further discussed later.

The fact that matapa is consumed with such regularity is a positive sign. Indeed, this evidence supports claims made by Smith and Eyzaguirre (2007) and Jansen Van Rensberg et al. (2007) about the important role leafy green vegetables might play in increasing vegetable consumption and mitigating malnutrition among the population in sub-Saharan Africa. It also shows that the vegetables being promoted by AKF(M) are culturally relevant, which is an important factor in increasing their consumption, as Krishna (2004) discusses.

Yet, as data from this study show, even matapa is not necessarily available at all times. Many plants used as sources of matapa are often only seasonally available and are affected by factors such as water availability and crop destruction by animals. However, moringa, being a fast-growing woody perennial, was one source of matapa leaves cited as being available almost year round, which underscores its potential in fighting malnutrition and food insecurity. Agbogidi and Ilondu (2012) provide much support for this, highlighting moringa's many uses both as a food and a medicinal product, its cultural relevance and its environmental adaptability and hardiness.

A further issue, as raised by both Uusiku et al. (2010) and Krebs-Smith and Scott (2001), is that it is necessary to consume a variety of fruit and vegetables in order to fully take advantage of the health-protecting properties of the micronutrients present in this food group – something which was clearly not happening among study participants. As such, the need to support the production and consumption of a range of vegetables in the Cabo Delgado region is clear.

As discussed earlier, AKF(M)'s intervention is promoting production of quite a range of vegetables, which is in line with the need to do so as highlighted by these authors – a strength of the program.

Factors affecting vegetable consumption

Given that the majority of study participants understood that vegetables are good for health and indeed that diet diversity in general is also healthful, it stands that other barriers to consuming a diverse diet (including vegetables) must be present.

The most common barrier faced by research participants to accessing adequate vegetables was found to be cost. This is in line with research conducted by Keatinge et al. (2011), who state: "In recent decades, prices of fruits and vegetables have risen faster than those of bread or rice and have rapidly grown beyond the reach of the poor so that fresh produce is no longer regularly purchased and consumed." FAO (2008), in Keatinge et al. (2011), found that high prices for staples in 2008 forced many low-income households worldwide to reduce diet diversity by restricting intake of "expensive" foods such as meat, fruits and vegetables and to eat fewer meals per day or to reduce portion sizes. This resulted in lower energy intake and increased levels of micronutrient deficiencies.

Destruction of crops by animals – in both machambas and home gardens – was also commonly referenced as an issue, particularly in the more rural villages. This too was a commonly cited barrier to vegetable access in other research (Musotsi, 2008; Yiridoe and Anchirinah, 2005).

Additionally, it is necessary to address the issue of nutrition-based knowledge. As discussed earlier, it was found that many research participants had at least a general understanding of the importance of consuming vegetables. However, there seemed to be a mixed understanding on the matter and on nutrition in general. In particular, when the scope is widened to consider more than just fruits and vegetables, it is evident that in many cases there is not a clear understanding of what is healthy and what is not.

Of particular interest is the fact that a number of individuals voiced the idea that while vegetables (primarily, but also other foods it seems) do indeed have many vitamins, they must be cooked in oil for them to be healthy. As one man said: "We only eat to fill our bellies. [These foods] we eat don't have any vitamins because to get the vitamins you have to use all things that are required. We only boil everything." Similarly, a woman stated: "All this food it gives us vitamins. The problem is that we don't prepare [it] as we are supposed to do" (that is, cooked in oil).

These quotes stand out as interesting in light of some of the potential issues surrounding the bioavailability of vegetable-based nutrients (de Pee and Bloem, 2007; de Pee et al., 1995). However, once again, understanding of the issue does not seem to be fully accurate. There is merit to the belief that vegetables are not nutritious at all unless they are cooked, or cooked in oil, given the positive effect oil has on absorption of fat-soluble pro-vitamin A. However, the perception that vegetables not cooked in oil have no nutritional value may be detrimental to efforts to increase vegetable consumption.

Unfortunately, the extent to which respondents' nutrition-based knowledge affects their nutrition practices cannot be determined from this present study due to the limited nature of the data collected on the matter. Regardless, it can be seen that there is room for improvement in this area.

Moreover, as discussed, much literature on the topic indicates that nutrition education is a crucial component of development initiatives aimed at improving health outcomes among the target population. Numerous authors have highlighted the importance of coupling interventions aimed at improving food production and/or access with those that promote good nutrition knowledge and practices (Faber and Laurie, 2010; Iannotti et al., 2009; Berti et al., 2004; Faber et al., 2002a, 2002b). Data from this study, combined with aforementioned literature, provide support for the nutrition education component of AKF(M)'s current FSI initiative.

Home gardening as an effective intervention in Cabo Delgado?

It has been proven that home gardening initiatives can be a successful component of broader nutrition intervention programs (DFID, 2014; Masset et al., 2012; Cabalda et al., 2011; Ecker et al., 2010; Keatinge et al., 2011; Talukder et al., 2010; Iannotti et al., 2009; Laurie and Faber, 2008; Musotsi et al., 2008; Yiridoe and Anchirinah, 2005; Eyzaguirre and Linares, 2004). However, their potential has yet to be fully investigated in the Cabo Delgado region.

In order for the home gardening project in question to succeed, the practice must be easily adopted by a large number of potential beneficiaries in the target population; barriers to starting and maintaining a home garden should not outweigh its potential benefits. Beneficiaries must also be amenable to adopting the practice in the first place. Additionally, in order for the intervention to be sustainable, it should be easily replicated and disseminated within the population without the need for extensive and continued intervention by AKF(M). Furthermore, once established, the home garden must be easy to maintain and sustain from season to season.

In discussing knowledge of home gardening with people in districts targeted by AKF(M)'s development intervention, even those without home gardens were found to be aware of the practice and to hold a generally favorable view of it. The primary way individuals learned of home gardening in the first instance seemed to be through observation of other members of their village who had a home garden. As one man put it: "They [home gardens] are not hidden, you can easily see one existing in someone's compound when you go pass by or go in."

In many cases, awareness of home gardening included a willingness to adopt it, as indicated by both community members and AKF(M) agriculture supervisors and facilitators. However, this did vary district to district; in one district no one had any desire to start a home garden at all, for some of the reasons previously discussed. Moreover, most individuals seem to believe that this is a practice that is possible for anyone to do. Other data from this study support this notion. Indeed, it does seem that home gardening is something that is accessible to most, including poorer members of society who are the most vulnerable to nutrition and food security–related stresses. Barriers to adoption that are related to socioeconomic status (e.g., access to inputs or money) did not appear to be particularly significant, as is discussed later.

These findings are in line with conclusions drawn by Faber and Laurie (2010) and Berti et al. (2004), who found that home gardening is an intervention most households can successfully adopt. Moreover, the fact that cost was found to be the number one barrier research participants face to accessing vegetables further validates home gardening as a means of increasing vegetable consumption, particularly as home gardens have proven to be a cost-effective manner of increasing access to vegetables (Keatinge et al., 2011; Faber and Laurie, 2010). As such, their relevance in this context is clear.

When asked why they adopted the practice, individuals with home gardens cited a number of hoped-for outcomes, including the desire to have better access to vegetables year-round (the most common motive by far), the possibility of improving their household's nutrition and of earning some income. However, the latter were much less common drivers than the issue of access.

One focus group member stated: "I started the home garden because I wanted to decrease the distance from the village to the farm when I want vegetables." A personal interviewee indicated that she and her husband started their home garden for the same reason: "We wanted to decrease the distance of going to the farm which is far to get the vegetables." Among individuals without home gardens, the same positive outcomes were cited as reasons that they might consider starting the practice themselves.

Barriers to starting and maintaining a home garden

Nevertheless, there are barriers to home gardening in this region that may pose a significant challenge for many individuals with gardens and prevent others from taking up the practice. In particular, individuals – both with and without home gardens – regularly raised concerns about issues pertaining to water access and animal control. This is not surprising given that the region is situated in a semi-arid environment and is prone to droughts, and many of the villages where research was conducted are in fact located in a national park with much wildlife.

The number one barrier cited by respondents was animal interference. According to numerous individuals, the primary culprits were chickens, goats and monkeys. Apparently, chickens eat seeds and seedlings before they have a chance to grow, and they destroy mulch soil cover; goats eat anything that has managed to grow and are known to destroy any fencing that has been put in place; while monkeys simply tend to "destroy the garden." The latter seems to occur with or without appropriate fencing.

Again, a number of other studies have found animal destruction of crops to negatively impact the maintenance of home gardens among study participants across Africa, from South Africa to Kenya to Ghana (Faber and Laurie, 2010; Musotsi, 2008; Yiridoe and Anchirinah, 2005). This was highlighted as an ongoing issue for participants in this current intervention in Cabo Delgado, as identified in a brief follow-up study conducted by AKF(M) in 2013 (AKF, 2013). Fencing is a possible solution to this problem, although access to capital to purchase the necessary materials may be a barrier unto itself, as found by Musotsi (2008). Additionally, as participants of the current study indicated, monkeys seemed to be a problem with or without fencing. In this instance, no obvious solution is immediately available, although it is possible that further research will uncover a solution.

Water access was the second-most frequently cited barrier, although the degree to which it affected production varied significantly from village to village. Iannotti et al. (2009) highlight similar problems related to home garden irrigation in their discussion of HKI's home gardening initiative in Bangladesh, as did Faber and Laurie (2010) in South Africa.

Discussion with AKF(M)'s current coordinator of monitoring and evaluation (2015) brought further light to water access as an issue in Cabo Delgado, which has been an ongoing concern since AKF's home gardening intervention commenced in 2011. During the dry season (August to November), it rains very little and a number of villages have very limited access to water above and beyond what is used to meet households' basic consumption needs (cooking, washing, etc.). He indicated that efforts were being made to promote the use of wastewater in watering home gardens, but that results have been mixed so far – perhaps this is an area for further investigation.

Lastly, access to seeds, finding sufficient time, insect control, theft and access to space were also mentioned but seemed to be much less common or pressing issues than either animal control or water access.

Home gardening's impact on nutrition

Above and beyond the likelihood that the target population will adopt home gardening, the intervention's ability to impact the nutrition of the population must also be considered. That is, does the intervention have the potential to achieve its aim of increasing micronutrient intake among the target population (i.e., women and children) through increased fruit and vegetable consumption?

At the time of this study it was too early to determine whether or not the home gardens in question were having a measurable impact on fruit and vegetable consumption; the intervention was only in its first season. However, it was found that early adopters of the practice were already experiencing a number of positive outcomes, many of which are similar to those achieved by other home gardening interventions. Some of these have the potential to directly and positively impact the nutritional status of household members.

In particular, access to vegetables seems to have been greatly enhanced in line with other home gardening studies (Keatinge et al., 2011; Faber and Laurie, 2010; Ecker et al., 2010; Ianotti et al., 2009). Study participants with home gardens were growing a wide range of produce including favorites such as cassava, beans, tomatoes, cabbage and pumpkins, among many others. Much of this produce was consumed within their households, a fact that alone indicates it may be possible for nutrition status to be improved by this intervention, because, as highlighted by various authors, access to food is one of the primary determinants of nutrition status (Foresight, 2011; Webb and Rogers, 2003; Sen, 1981).

Many participants also had enough produce to sell some of their harvest, thereby generating income for their household, another commonly cited benefit that has great potential to positively impact nutrition at the household level (AKF, 2013; Keatinge et al., 2011; Iannotti et al., 2009). As discussed by many, a household's ability to access adequate food often depends significantly on its income (Foresight, 2011; Webb and Rogers, 2003; Sen, 1981).

Of note is that the majority of research participants who had home gardens were women, many of whom indicated they directly control any income generated from their gardens. This may have an additional positive impact on nutrition. As much previous research has shown, income that is controlled by women rather than men is much more likely to be invested in improving the quality of life, health and nutrition of household members, children in particular (Iannotti et al., 2009; Ruel, 2008).

Lastly, it is clear that the benefits of home gardening experienced thus far are in line with the "potential benefits" many individuals cited when asked why they started (or would start) a home garden in the first place. This indicates peoples' expectations are being met, or can be met − a fact that may greatly contribute to the success and sustainability of this intervention.

AKF(M) FSI intervention: developments and areas for further research

A number of questions remain unanswered following this initial investigation into home gardening practices and their impact on the food security and nutrition of the rural population of Cabo Delgado. First and foremost, the effect the home gardens in question are having on the nutrition status of people in the Cabo Delgado region must be evaluated (1) after the project has had time to become established to a greater extent and (2) during the region's lean season. As it is lean-season nutritional stress that this home gardening project ultimately aims to alleviate, the true impact the practice has on mitigating malnutrition must be investigated during the time when households have the most difficulty obtaining sufficient, nutritious food to meet their basic needs.

The question of nutrition-based knowledge and practices should also be looked at in greater depth. In particular, it would be useful to discuss questions similar to those investigated in this study, but with the two groups (individuals with and without home gardens) being interviewed at separate times (which did not happen in this study). This would give a much clearer picture of how home gardening practices are affecting nutrition in the region.

Lastly, the relationship between socio-economic status and home gardening practices remains unclear. This is another potentially important factor that merits further investigation.

A full and in-depth evaluation of AKF's FSI project in Cabo Delgado will be undertaken in January 2016 and may answer some of these questions. That it will take place during the lean season is important, allowing the study to provide as accurate a picture as possible of the food security situation in Cabo Delgado.

An interim study was undertaken in January 2013, which did provide some illuminating data (AKF, 2013). Of surveyed households in the program area, 26.7% had established and maintained a home garden. The proportion did not vary greatly by livelihood zone and type of household. There is clearly much room for improvement in this, but encouragingly, progress has been made.

This study also found that home gardens not only improved households' access to nutritious foods, but also enabled them to generate income from the sale of vegetables, particularly during non-lean seasons. In light of the fact that it was also found that home gardens are generally maintained and controlled by women – as in Attorp's original study (2011) – this may have a compound, positive effect on household nutrition given that, as discussed, income controlled by women rather than men is much more likely to be invested in improving the quality of life and health of household members (Iannotti et al., 2009; Ruel, 2008).

A last positive finding to note was that home garden owners did not feel the work required to maintain a garden was excessive. They mostly reported being able to quite easily find time to work in the garden. Thus it can be seen that time constraints, at least in this case, do not appear to be a barrier to implementing the practice.

However, as in Attorp's study of the program, access to water and problems with animal damage were identified as significant barriers for individuals wanting to start a home garden and for those already maintaining one. Indeed, some study participants who had started home gardens had been forced to abandon the practice because of these problems. As a result, this has been identified as an area of priority for research during the remainder of the FSI (AKF, 2013). Results of the 2016 study should provide insight as to whether any progress has been made on this front since 2013.

With regard to water access, during the rainy season there is a need to maximize water capture and retention in order to reduce the impact of any drought stress. It is equally important to maximize transpiration and minimize water loss through soil evaporation. These outcomes are attainable with the approach to production intensification with Conservation Agriculture (CA), in which soil disturbance is minimized or avoided when possible and the soil is kept covered with mulch. FSI's agricultural intensification strategy is based on the adoption and spread of CA across all the target districts, and there is evidence that CA has enabled not only greater yields but also reduced runoff and erosion and improved soil water retention (Jat et al., 2014; Kassam, 2020). During the dry season, reducing plant water requirements would help extend the limited water supplies. Again, there is good evidence from the agricultural intensification component of FSI that water requirements of vegetable field crops are reduced by some 30% or more because of the use of soil mulch cover and greater soil water retention. Also, there is a significant increase in water productivity with CA resulting from better water capture and more water being used for growth. Thus there is a need to conduct adaptive tests to assess the usefulness of applying CA practices of no-till and soil mulch cover to home gardens.

Conclusions

Globally, micronutrient deficiencies have a significant negative impact on the well-being of a large percentage of the population – children in particular. Fruits and vegetables are considered a significant source of micronutrients, and it is suggested that consuming a variety of these daily can help mitigate a number of risk factors for both micronutrient deficiencies and disease.

Home gardens are common in many developing countries and play an important role in diversifying diets and increasing food security for both urban and rural households. They are significantly associated with increased consumption of fruits and vegetables and thus increased micronutrient intake. Home gardening initiatives may appreciably contribute to improving micronutrient consumption among vulnerable populations.

It is clear there is a significant malnutrition problem in Cabo Delgado, Mozambique. In particular, for many participants of this study, the inability to obtain year-round access to sufficient vegetables appears to be a major barrier to achieving good nutrition. Issues pertaining to nutrition-based knowledge and practices and the impact these have on nutrition status merit further investigation.

AKF(M)'s home gardening intervention is well suited to mitigate some of these problems, particularly when viewed in the context of its greater FSI initiative (e.g., when coupled with relevant nutrition education programs). Benefits to women and children may be particularly strong.

The home gardening practices being promoted by the organization seem to be well received by the target population. They also appear to be accessible to almost all individuals, regardless of status or socioeconomic position. Note, however, that the matter of socioeconomic status may need to be examined more closely, particularly as it relates to nutrition practices and health.

Plants being promoted by AKF(M)'s agriculture department are suitable in terms of their potential ability to supply vegetables year-round in their favorable nutrient profile and cultural relevance. For the bioavailability of certain nutrients in plants, they must not be overlooked, but it is likely that this issue may addressed, at least in part, through education about proper cooking methods.

In short, the malnutrition problem faced by the population of Cabo Delgado Province is broad and multifaceted. Results from this study show that home gardens are a feasible means of increasing individuals' fruit and vegetable consumption (and by extension, micronutrient consumption). Accordingly, there is reason to believe appropriate interventions, such as AKF(M)'s home gardening project, may have a significant positive impact on malnutrition in the region.

Acknowledgments

Thank you to the Aga Khan Foundation (AKF) (Mozambique) and Ms. Faiza Janmohamed, CEO of AKF Mozambique, for providing the opportunity to undertake this study. Thank you also to all the AKF staff members based in Pemba at the time of the study, the Monitoring and Evaluation Team in particular, who helped make the research possible.

References

Aga Khan Foundation (AKF), 2013. Food security study report for the Food Security and Incomes (FSI) project funded by the Canadian International Development Agency (CIDA). Aga Khan Foundation (Internal Document).

Aga Khan Foundation Canada (AKCF), 2011. Enhancing food security and incomes in Northern Mozambique initiative: Project implementation plan. Aga Khan Foundation Canada (Internal Document).

Aga Khan Foundation (Mozambique) (AKF(M)), 2009. Household survey report: Quality of life assessment, Cabo Delgado Mozambique. Aga Khan Foundation (Mozambique) (Internal Document).

Agbogidi, O.M. and Ilondu, E.M., 2012. Moringa Oleifera LAM: Its potentials as a food security and rural medicinal item. *Journal of Bio Innovation*, 1, pp. 156–167.

Attorp, A., 2011. *Home gardens and malnutrition among women and children in Cabo Delgado, Mozambique.* MSc Thesis, University of Reading, UK.

Bhattacharjee, L., Phithayaphone, S. and Nandi, B.K., 2006. *Home gardens key to nutritional well-being.* Available at: www.fao.org/3/a-ag101e.pdf [Accessed 11 July 2015].

Berti, P.R., Krasevec, J. and FitzGerald, S., 2004. A review of the effectiveness of agriculture interventions in improving nutrition outcomes. *Public Health Nutrition*, 7, pp. 599–609.

Black, R.E., Allen, L.H., Bhutta, Z.A., Coulfield, L.E., de Onis, M., Ezzati, M., Mathers, C. et al., 2008. Maternal and child undernutrition: Global and regional exposures and health consequences. *Lancet*, 371, pp. 243–260.

Bloem, M.W., Huq, N., Gorstein, J., Burger, S., Khan, T., Baker, S. and Davidson, F., 1996. Production of fruits and vegetables at the homestead is an important source of vitamin A among women in rural Bangladesh. *European Journal of Clinical Nutrition*, 50(S3), pp. S6–S67.

Cabalda, A.B., Rayco-Solon, P., Solon, J.A. and Solon, F.S., 2011. Home gardening is associated with Filipino preschool children's dietary diversity. *Journal of the American Dietetic Association*, 111, pp. 711–715.

Concern Worldwide, 2009. *The nexus between agriculture and heath: Concern Worldwide Programs*. Dublin, Ireland: Concern Worldwide.

Demographic and Health Surveys Program, 2011. *Demographic health survey for mozambique*. Washington, DC: USAID.

Department for International Development (UK), 2014. *Can agricultural interventions promote nutrition? Agriculture and nutrition evidence paper*. London, UK: Department for International Development.

De Pee, S. and Bloem, M.W., 2007. The bioavailability of (pro)vitamin A carotenoids and maximizing the contribution of homestead food production to combating vitamin A deficiency. *International Journal for Vitamin and Nutrition Research*, 77, pp. 182–192.

De Pee, S., West, C.E., Muhilal, Karyadi, D. and Hautvast, J., 1995. Lack of improvement in vitamin A status with increased consumption of dark-green leafy vegetables. *The Lancet*, 346, pp. 75–81.

Diana, R., Khomsan, A., Sukandar, D. and Riyadi, H., 2014. Nutrition extension and home garden intervention in Posyandu: Impact on nutrition knowledge, vegetable consumption and intake of vitamin A. *Pakistan Journal of Nutrition*, 13, pp. 88–92.

Ecker, O., Weinberger, K. and Qaim, M., 2010. Patterns and determinants of dietary micronutrient deficiencies in rural areas of East Africa. *African Journal of Agriculture and Resource Economics*, 4, pp. 175–194.

The Edible Schoolyard Project, 2015. *School lunch reform*. Available at: http://edibleschool yard.org/node/364/#sli [Accessed 21 November 2015].

Eyzaguirre, P.B. and Linares, O.F., 2004. *Home gardens and agrobiodiversity*. Washington, DC: Smithsonian Books.

Faber, M. and Laurie, S., 2010. A home gardening approach developed in South Africa to address vitamin A deficiency. In Food and Agriculture Organization, ed. (2011). *Combating micronutrient deficiencies: Food-based approaches*. Oxfordshire, UK and Rome, Italy: CAB International and Food and Agriculture Organization, pp. 163–182.

Faber, M., Phungula, M., Venter, S.L., Dhansay, M.A. and Spinnler Benade, A.J., 2002a. Home gardens focusing on the production of yellow and dark-green leafy vegetables increase the serum retinol concentrations of 2–5 year old children in South Africa. *American Journal of Clinical Nutrition*, 76, pp. 1048–1054.

Faber, M., Venter, S. and Spinnler Benade, A.J., 2002b. Increased vitamin A intake in children aged 2–5 years through targeted home-gardens in a rural South African community. *Public Health Nutrition*, 5, pp. 11–16.

FAO, 2008. *The state of food insecurity in the world 2008: High food prices and food security – threats and opportunities*. Rome, FAO. Available at: http://www.fao.org/3/i0291e/i0291e00.htm [Accessed 9 April 2020].

FAO, IFAD and WFP, 2014. *The state of food insecurity in the World 2014: Strengthening the enabling environment for food security and nutrition*. Rome, FAO.

FAO, IFAD and WFP, 2015. The state of food insecurity in the world 2015. *Meeting the 2015 International Hunger Targets: Taking Stock of Uneven Progress*. Rome, FAO, pp. 930–931.

Foresight, 2011. The future of food and farming: Challenges and choices for global sustainability. *Final Project Report*. London, UK: The Government Office for Science.

Harris, J., 2011. *Agriculture, nutrition and health essentials for non-specialist development professionals*. Washington, DC, USA: International Food Policy Research Institute.

Hotz, C., Loechl, C., de Brauw, A., Euzenou, P., Gilligan, D., Moursi, M., Munhaua, B. et al., 2012. A large-scale intervention to introduce orange sweet potato in rural Mozambique increase vitamin A intake among children and women. *British Journal of Nutrition*, 108, pp. 163–176.

Iannotti, L., Cunningham, K. and Ruel, M., 2009. *Improving diet quality and micronutrient nutrition: Homestead food production in Bangladesh.* Washington, DC: International Food Policy Research Institute.

International Bank for Reconstruction and Development, 2009. *Implementing agriculture for development.* Washington, DC, USA: The World Bank.

Jansen Van Rensberg, W.S., Van Averbeke, W., Slabbert, R., Faber, M., Van Jaarsveld, P., Van Heeden, I., Wenhold, F. and Oelofse, A., 2007. African leafy vegetables in South Africa, In Uusiku, N.P., Oelofse, A., Duodu, K.G., Bester, M. and Faber, M. (2010) Nutritional value of leafy vegetables of sub-Saharan Africa and their potential contribution to human health: A review. *Journal of Food Composition and Analysis*, 23, pp. 499–509.

Jat, R., Sahrawat, K. and Kassam, A., 2014. *Conservation agriculture: Global prospects and challenges.* Wallingford, UK: CABI International.

Kassam, A.H. 2020. *Advances in conservation agriculture volume 2: Practice and benefits.* Cambridge, UK: Burleigh Dodds.

Keatinge, J.D.H., Yang, R.Y., Hughes, J.d'A., Easdown, W.J. and Holmer, R., 2011. The importance of vegetables in ensuring both food and nutritional security in attainment of the Millennium Development Goals. *Food Security*, 3, pp. 491–501.

Krebs-Smith, S.M. and Scott, K.L., 2001. Choose a variety of fruit and vegetables daily: Understanding the complexities. *Journal of Nutrition*, 131, pp. 487S–501S.

Krishna, G.C., 2004. Home gardening as a household nutrient garden. In R. Gautam, B. Sthapit and P.K. Shrestha, eds. (2006). *Home gardens in Nepal: Proceeding of a workshop on Enhancing the contribution of home garden to on-farm management of plant genetic resources and to improve the livelihoods of Nepalese farmers: Lessons learned and policy implications*, 6–7 August 2004. Pokhara, Nepal, LI-BIRD, Bioversity International and SDC.

Laurie, S.M. and Faber, M., 2008. Integrated community-based growth monitoring and vegetable gardens focusing on crops rich in β-carotene: Project evaluation in a rural community in the Eastern Cape, South Africa. *Journal of the Science of Food and Agriculture*, 88, pp. 2093–2101.

Masset, E., Haddad, L., Cornelius, A. and Isaza-Castro, J., 2012. Effectiveness of agricultural interventions that aim to improve nutritional status of children: Systematic review, *BMJ.* Available at: www.bmj.com/content/bmj/344/bmj.d8222.full.pdf [Accessed 30 June 2015].

Musotsi, A.A., Sigot, A.J. and Onyango, M., 2008. The role of home gardening in household food security in Butere Division of Western Kenya. *African Journal of Food, Agriculture, Nutrition and Development*, 8, pp. 376–390.

Nutrition and Consumer Protection Division, 2010. *A new deal for school gardens.* Rome, Italy: Food and Agriculture Organization of the United Nations.

Rauzon, S., Wang, M., Studer, N. and Crawford, P., 2010. *Changing student's knowledge, attitudes and behavior in relation to food: An evaluation of the school lunch initiative.* Berkeley, USA: Dr. Robert C. and Veronica Atkins Center for Weight and Health, University of California at Berkeley.

Ruel, M., 2008. Addressing the underlying determinants of undernutrition: Examples of successful integration of nutrition in poverty-reduction and agriculture Strategies. *SCN News*, 36, pp. 21–29.

Selepe, M. and Hendriks, S., 2014. The impact of home gardens on pre-schoolers nutrition in Eatonside in the Vaal Triangle, South Africa. *African Journal of Hospitality, Tourism and Leisure*, 3. Available at: www.ajhtl.com/uploads/7/1/6/3/7163688/article_17_vol.3_2_july_14.pdf [Accessed 11 July 2015].

Sen, A., 1981. *Poverty and famines: An essay on entitlement and deprivation.* Oxford, UK: Oxford University Press.

Smith, I.F. and Eyzaguirre, P., 2007. African leafy vegetables: Their role in the World Health Organization's global fruit and vegetable initiative. In N.P. Uusiku, A. Oelofse, K.G. Duodu, M. Bester and M. Faber, eds. (2010). Nutritional value of leafy vegetables of sub-Saharan Africa and their potential contribution to human health: A review. *Journal of Food Composition and Analysis*, 23, pp. 499–509.

Soil for Life, 2013. Available at: http://soilforlife.co.za/why-we-do-it [Accessed 21 November 2015].

Sthapit, B., Gautam, R. and Eyzaguirre, P., 2004. The value of home gardens to small farmers. In R. Gautam, B. Sthapit and P.K. Shrestha, eds. (2006). *Home Gardens in Nepal: Proceeding of a workshop on Enhancing the contribution of home garden to on-farm management of plant genetic resources and to improve the livelihoods of Nepalese farmers: Lessons learned and policy implications,* 6–7 August 2004. Pokhara, Nepal, LI-BIRD, Bioversity International and SDC.

Sullivan, M., 2013. Seasonality: The missing piece of the undernutrition puzzle? *ACF International.* Available at: www.cmamforum.org/Pool/Resources/Seasonality-ACF-2013.pdf [Accessed 1 August 2015].

Talukder, A., Haselow, N.J., Osei, A.K., Villate, E., Reario, D., Kroeun, H., SokHoing, L., Uddin, A., Dhunge, S. and Quinn, V., 2010. Homestead food production model contributes to improved household food security and nutrition status of young children and women in poor populations-lessons learned from scaling-up programs in Asia (Bangladesh, Cambodia, Nepal and Philippines). *Field Actions Science Reports.* [Online], Special Issue 1. Available at: http://factsreports.revues.org/404 [Accessed 11 July 2015].

Thompson, B. and Amoroso, L. eds., 2010. *Combating micronutrient deficiencies: Food-based approaches.* Rome: FAO.

UNICEF, 2011. *Mozambique: Child survival-nutrition.* Available at: www.unicef.org/mozambique/child_survival_4895.html [Accessed 1 August 2015].

United Nations, 2010. *2010 human development index.* Available at: http://hdr.undp.org/en/media/Lets-Talk-HD-HDI_2010.pdf [Accessed 1 August 2015].

United Nations, 2014. *Human development report: Sustaining human progress: Reducing vulnerabilities and building resilience: Explanatory notes on the 2014 human development report composite indices: Mozambique.* Available at: http://hdr.undp.org/sites/all/themes/hdr_theme/country-notes/MOZ.pdf [Accessed 1 August 2015].

Uusiku, N.P., Oelofse, A., Duodu, K.G., Bester, M. and Faber, M., 2010. Nutritional value of leafy vegetables of sub-Saharan Africa and their potential contribution to human health: A review. *Journal of Food Composition and Analysis*, 23, pp. 499–509.

von Grebmer, K., Saltzman, A., Birol, E., Wiesmann, D., Prasai, N., Yin, S., Yohannes, Y., Menon, P., Thompson, J. and Sonntag, A., 2014. *2014 global hunger index: The challenge of hidden hunger.* Bonn, Germany, Washington, DC, USA, and Dublin, Ireland: International Food Policy Research Institute, Concern Worldwide and Welthungerhilfe.

Webb, P. and Rogers, B., 2003. *Addressing the "in" in food insecurity.* Washington, DC, USA: USAID Office of Food for Peace.

West, K.P., Jr., 2003. Vitamin A deficiency disorders in children and women. *Food and Nutrition Bulletin*, 24, pp. S78–S90.

World Bank Group, 2015. *Mozambique data and statistics.* Available at: http://data.worldbank.org/country/mozambique [Accessed 21 November 2015].

World Health Organization, 2015. *Food security.* Available at: www.who.int/trade/glossary/story028/en/ [Accessed 1 August 2015].

Yiridoe, E.K. and Anchirinah, V.M., 2005. Garden production systems and food security in Ghana: Characteristics of traditional knowledge and management systems. *Renewable Agriculture and Food Systems*, 20, pp. 168–180.

6 Home garden experiences in Costa Rica

Helga Blanco-Metzler and Alex Diaz Porras

Some of the factors that characterize the northern region of Costa Rica are (1) an increase in deforestation; (2) the lack of arable land; (3) damage to systems caused by the wind; (4) the absence of sustainable production alternatives; (5) dependence on imported supplies which make production more expensive and end up polluting the environment, the soils and the water; (6) lack of protein in the diet of inhabitants; (7) low consumption of fruits and vegetables; (8) deterioration of soils, native flora and fauna; and (9) the migration of farmers to the big cities to look for jobs. All of these factors directly affect the small farmer by limiting the potential value gained from working their land and limiting their ability to compete in the national market.

Thus by opening the market to new timber species and to the production of farming products like coffee and vegetables using a minimum of pesticides, the goal is to strengthen and diversify the agricultural holdings so that they are competitive and sustainable within the national economy. Also, by developing production systems in organic farming that aim to produce a sufficient quantity of healthy food while minimizing the negative impact on the environment, the objective is to stimulate self-sufficiency of the farm and the country through a better use of agricultural resources and agro-industrial and livestock waste. Further, another goal was to have all the rural families involved in farming activities and to promote production diversity.

Traditional systems of soil use in Costa Rica, except shade farming of coffee and cacao, do not consider mixing forest species in with agricultural crops. In general, our farmers establish their agricultural holdings based on monoculture, mixed home gardens and annual crops put in between fruit trees, without considering the introduction of a forest specie which, if suitable, could become a medium- or long-term additional income.

The agroforestry and agro-silvopastoral systems can help reduce soil use problems because factors that favor the production system, such as the increase of soil nutrient recycling, the improvement of the soil quality, better protection from erosion, the increase of the soil's capacity to store water and the reduction of problems with insects that damage the crops, all enhance the diversity of the system and number of parasitoids and predators.

In the past, forestry research of native species was limited because it was directly carried out in forests. However, research in the last decade has proved it is feasible to include native species in plantations if the appropriate techniques to prepare the soil for pest control and for silvicultural treatment are followed.

In spite of the low international coffee prices, coffee production in Costa Rica is still an important commercial activity. Its contribution to the gross national internal product during the last few years has been around 15% and it represents approximately 4% of the total exports. About 91% of the coffee production is concentrated in small or medium size farms (Instituto del Café de Costa Rica (ICAFE), 2009, 2011); this provides employment and food for more than 50,000 farming families, 145 beneficiaries, 55 roaster companies and 60 exporting companies (Instituto Nacional de Estadística y Censos, 2009) and during the harvest for more than 300,000 people. The coffee production in Costa Rica focuses on the quality of the bean and not the volume, so different varieties of beans are grown but they are not as productive by volume as the ones produced in Brazil (Robusta) (ICAFE, 2011).

The Montes de Oro region, in the Province of Puntarenas, Costa Rica, is a marginal agricultural area with coffee production as its main activity, although during the last few years there has been a substantial migration of farmers to cities in search of better job opportunities. The region faces a number of social and economic problems, worsened by the reduction of forest areas, the increase in soil erosion, the absence of sustainable land production alternatives and a heavy dependence to imported pesticides. Therefore, there have been efforts to protect the region's biodiversity and to mitigate the negative environmental effects through maintenance or implementation of organic coffee production systems and through integrating ecological, social and economic changes to offer more sustainable and profitable production alternatives (Blanco-Metzler and Diaz-Porras, 2003). Such efforts were focused on seven components: (1) associated crops and commercialization of vegetables, (2) establishment of shade trees using fruit and native forest trees, (3) establishment of windbreaks, (4) coffee fertilization, (5) studies on bird diversity, (6) improved coffee processing systems and (7) introduction of animal protein to the diet of the farming families.

Production and marketing of vegetables

A survey was carried out to identify the crops more frequently intercropped with coffee. According to the survey, the main crops associated with coffee plantations were tomato, sweet pepper, dry and green beans and corn. These crops are grown when the coffee trees are pruned because it allows the growth of vegetables for one to two years. Farmers are said to prefer these crops as they are well known, have a safe market and are easy and fast to produce. However, crop choices also have to do with tradition and the lack of knowledge about other promising crops and improved varieties. Farmers were trained on new technologies used in organic vegetable production (greenhouses, seedling production, pest control and improved varieties).

Seedling production

The objective of this program was to familiarize the farmers with modern techniques in the germination and handling of seedlings. Farmers participated in all the research phases: substratum elaboration, filling of trays, sowing, monitoring germination and measurement of the growth of seedlings, both by aerial measurement and by measurement of roots. Some of the practices are:

- Greenhouse with antivirus net that will protect it from the insects that transmit viruses and diseases. Farmers also learned the advantages of having a double door; sanitizing their shoes, appropriate weeding with cleaning practices where weeds are not supposed to be thrown back on the floor but into a box
- Use of trays (plastic and Styrofoam) in the production of seedlings
- Use of appropriate substratum. Seven substrata were used (Photo 6.1, Table 6.1)

Photo 6.1 Seedlings grown in different substrata evaluated, Montes de Oro, 2003

Table 6.1 Evaluation of two substrata and the types of measurements done by farmers

	Germination (%)	Number of leaves	Number of true leaves	Height (cm)
Scallion				
Peat moss + soil	90.8	1		2.5
Bocashi + soil	91.8	1		2.5
Lombricompost				
Fermented soil	3	1		0.5
Compost + soil	80.6	1		1.5
Lombricompost + soil + parchment	37.8	1		2.5
Coffee dregs + soil + rice husks + lime	93.8	1		2.5
Iceberg lettuce				
Peat moss + soil	99	2	1	2.5
Bocashi + soil	97	2	2	3.5
Lombricompost				
Fermented soil	7.1	1	0	0.3
Compost + soil	8.1	3	0	0.5
Lombricompost + soil + parchment	92.8	2	2	1.8
Coffee dregs + soil + rice husks + lime	91.8	2	2	2.9

Source: Prepared by authors.

- Sowing of different vegetable seeds
- Introduction of vegetable seeds with high genetic value
- Storage of seeds
- Incidence of various types of risks and frequency.

Introduction of new horticultural crop varieties

When the greenhouse stage finished, the different varieties of vegetables were sown. In order for people to see the development of the crops, a farm located in a high traffic area was selected. To give an example, the farmers were used to seeing one single variety of lettuce, but in this project six varieties of lettuce were evaluated. Hence, it was possible to learn about other types of lettuces such as iceberg, Boston and red lettuce and to evaluate their adaptability under the weather conditions of the area. At the same time, palatability tests were carried out to verify its commercial potential.

Fertilization and pest management in vegetables

The elaboration of *bioles* (a type of organic fertilizer) and insect repellents made out of molasses as an extractor of secondary metabolites and medicinal plants was taught through participative training (Photos 6.2a, 6.2b and 6.2c).

(a)

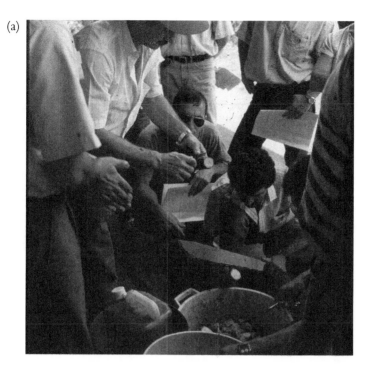

Photo 6.2a Farmers chopping fruit and medicinal plants to make bioles

(b)

Photo 6.2b Addition of molasses to chopped fruits and medicinal plants

(c)

Photo 6.2c Oxygen extraction during the elaboration of bioles and insect repellents

Products commercialization

One of the elements that affects the economy of farmers is the lack of organization among them when they try to sell their products in the market. Each
time they harvest their crops, the ones that have the resources move their products to the wholesale marketplace that is located in the capital of the country. Thus there are high transportation costs to the farmer. The farmers that
cannot afford transportation costs are forced to sell their harvest to brokers
that pay lower prices for their products. Therefore, an inter-institutional effort
was made with the purpose of organizing a vegetable gathering center for the
region through different visits to the communities and also through farmers'
surveys. The main market problems identified in the survey were high transportation costs (43%), brokers (15%), lack of markets (13%), low and unstable
prices (8%), bad roads (6%), poor farmers' organization (5%) and other factors
(10%). In this case, farmers were asked to suggest ideas to improve the crop
market. They are aware that a change in their attitudes is necessary and the
fact that they should organize themselves to access safer markets, obtain better
prices and reduce production costs. However, they pointed out that there is a

need for more efforts from the local authorities to help them and also for more support to their activity as farmers.

Establishment of shade plants

The use of shade trees benefits coffee plantations in various ways: they enhance fauna diversity and help improve the ecological conditions of the production unit, and they also produce goods of immediate use (Benzing, 2001). Shade plants were produced in the Cedral Farmers' Association nursery. The production of nursery trees was low quality. For that reason, field training was included. In this training, substratum mixture tests were carried out for the filling of bags in which a third part of the mixture was made of organic substances such as compost, bocashi and lombricompost. Also, they were taught about the right way to fill bags and the best collocation of those bags on the threshing floors.

A total of 8,400 shade trees were distributed among farmers of Montes de Oro for 77.6 ha of coffee with shade. The tree species that were planted were: *Leucaena, Gliricidia sepium, Albizia adinocephala, Erythrina poeppigiana, Cassia* and *Alnus jorullensis*; of this number, 1,934 were fruit trees (avocado, various citrus species, macadamia and soursop). The aim of the inclusion of fruit trees in the coffee plantations and other production systems was to provide the farmers with a second income source by means of the commercialization of fruits. In that way, they could see it as a compensation for the days they had economic problems because of the lower prices of coffee.

Establishment of windbreaks

Windbreaks are formed by planting one or more lines of trees or foliage, of the same or of different plant species and of varying heights. They are planted parallel and perpendicular to predominant wind. The use of windbreaks reduces eolian erosion; protects crops, animals and water springs; and keeps pastures from drying out during summer (Photo 6.3). A total of 46,000 trees were planted for 37,000 m of windbreaks, with 367 ha of land protected five years later, when maximum growth of the trees was expected. Different arrangements of the following tree species were used: *Cassuarina equisetifolia, Eucalyptus* spp., *Eugenia jambos, Coutorea latiflora, Cupressus lusitanical* and tubu.

Coffee fertilization

With the aim of reducing the high costs of fertilization without affecting grain quality and production, four fertilization alternatives were evaluated, together with a generalized application of Bocashi (5 ton/ha); two annual applications of a physical mixture of nitrogen (N), potassium (K) and boron (B) at a cost of USD 60/ha per year; two annual applications of a physical mixture of N, K, magnesium (Mg) and B at a cost of USD 66/ha per year; two annual

Photo 6.3 Windbreak of *Eucalyptus deglupta* and *Casuarina* spp.

applications of a physical mixture of N, K, at a cost of USD 47/ha per year; two annual applications of a physical mixture of the formula 18-5-15-6-2 (600 kg/ ha per year) at a cost of USD 100/ha per year. All of the fertilizer alternatives resulted in higher productions than the conventional fertilizer scheme account- ing for an increase of 8%–10% yield for alternatives 1 and 3 and an increase of 18% yield for alternative 2. In conclusion, it is possible to reduce production costs without negatively affecting yield.

Bird diversity

A study was carried out to evaluate bird diversity and abundance in two areas of Montes de Oro. The first area (1,300 masl (meters above sea level)) was located in the very humid premontane life zone, dominated by coffee plantations, cattle farms, secondary forest and border of harvested forests. The second area (1,000–1,450 masl) corresponded to the montane life zone. A total of 151 bird species was found in the study area (Stiles, 1991). Ten species were exclusive to the border area between the premontane and montane forest, i.e., they were not present in coffee farms. From the 141 species found in the zone where coffee is grown (premontane forest), 28 species were never registered in cof- fee plantations, while the rest (80%) were observed in coffee and surrounding farms. The 28 species not registered in coffee plantations were insectivorous and frugivorous birds, dependent on the forest for feeding, or that occasionally come out the forest but needed native fruits for their nourishment (Stiles and Skutch, 1989; Fogden, 2000).

The number of birds found in coffee farms was similar to that of the sur- rounding zone. However, bird diversity was higher in the surrounding areas than in the coffee farms. Hence, we decided to plant native fruit trees within the coffee plantations to attract bird populations. The recommended tree spe- cies were *Citharexylum caudatum* (Verbenaceae), *Ficus pertusa* (Moraceae), *Trich- ilia havanensis* (Meliaceae), *Ocotea* and *Nectandra* spp. (Lauraceae), *Conostegia xalapensis* (Melastomataceae), *Dendropanax arbore* (Araliaceae) and *Sorocea tro- phoides* (Moraceae). It was also suggested that increasing the number of shade trees within the coffee plantations would increase the insect population for insectivorous birds. After being presented with the information about bird diversity and then planting trees to increase feeding sources within the coffee crops, the farmers were able to diversify their products and also improved sales of ground coffee by 15%.

Coffee processing

Multiple efforts had been made to improve the production of coffee in the field. However, it was necessary to address weaknesses in the processing of the final grain and in the differentiation between organic coffee and conven- tional coffee. A Compact Ecological Processing Unit (UCBE) was purchased to process the organic coffee (Photo 6.4). This unit reduces the water required

Photo 6.4 Compact ecological processing unit (UCBE) purchased to process organic coffee

for processing from 800 L to only 11 L (for 258 kg of coffee), also reducing contamination and the cost of treatment of residual water. In addition, there is a significant saving in energy, since there is no need to use the main plant to process the small amounts of coffee produced at the beginning and at the end of the harvesting season.

The use of the new unit also reduced the time from depulping to storage from ten days in conventional processing to three days, which increases the grain yield and quality. All these improvements resulted in the Coopemontes de Oro processing plant being granted ISO 14000 certification, which implies an improvement regarding environmental quality.

Farmers were receptive to the improvements suggested. From a total of 250 farmers from Montes de Oro, Puntarenas, 10% went into organic farming and are certified by ECOLOGICA. Their production was sold at USD 200/100 kg of roasted coffee in 2006; 80% sell their coffee as fair trade with a price of USD 131/100 kg; the rest of the producers are conventional, and their production is paid at USD 80/100 kg.

Introduction of protein sources of animal origin

Pasture hens were introduced in order to increase the consumption of animal protein through the availability of meat and eggs (Photo 6.5). At the same time gandul (*Cajanus cajan*) was sown in the areas surrounding the henhouses as a way to reduce the dependency and the cost of commercial chicken feed. Both animals and humans could eat this grain. Annatto (*Bixa orellana*), plantain (*Musa* spp.) corn, sweet potato and beans were also sown and the surplus egg production was sold in the community.

Photo 6.5 Pasture hens and plants sown for feed, Caño Negro, Alajuela, Costa Rica, 2011

(a)

(b)

Photos 6.6a and 6.6b Care network organization for women involved in vegetable and tree production, Caño Negro, Alajuela, Costa Rica, 2011

Care network

A care network for women and children was organized with the aim of giving women equal opportunities to participate in training and in the fieldwork production of trees and vegetables. A physical space was set up with a roof, chairs, table and some hammocks. This care network was led by women in advanced stages of pregnancy, by women who were breastfeeding, or by women who did not have the physical ability for the field work. These women were in charge of feeding and taking care of children. Moreover, children were equipped with sheets of paper and colored pencils to facilitate their entertainment while their mothers were at work (Photos 6.6a and 6.6b).

Acknowledgments

This work could not be done without the aid of Allan González, Johel Chávez, Luis Alpízar, Farmers Association of Cedral, Víctor Julio Arce and the financial support from the Dutch Government; Fundecooperación and CONARE Costa Rica.

References

Benzing, A., 2001. *Agricultura orgánica: fundamentos para la región andina*. Villingen-Schwenningen, Germany: Necker-Verlag, pp. 340–350.

Blanco-Metzler, H. and Diaz-Porras, A., 2003. *Organización de un modelo agroforestal sostenible en fincas de pequeños productores de Montes de Oro, Puntarenas'*. San José, Costa Rica: Universidad de Costa Rica/Ministerio de Agricultura y Ganadería, 57 p.

Fogden, M.P., 2000. Birds of the monteverde area. In *Monteverde: Ecology and conservation of a tropical cloud forest*. New York: Oxford University Press, pp. 541–552.

Instituto del Café de Costa Rica (ICAFE), 2009. Informe sobre la Actividad Cafetalera de Costa Rica. *Costa Rica*, 96 p.

Instituto del Café de Costa Rica (ICAFE), 2011. Guía Técnica para el Cultivo del Café. *ICAFE–CICAFE, Costa Rica*, 72 p.

Instituto Nacional de Estadística y Censos (INEC), 2009. Resultados de Censo cafetalero. *Costa Rica*. Consultado el 4 de octubre 2014. Disponible en. Available at: www.inec.go.cr/Web/Home/GeneradorPagina.aspx

Stiles, F.G., 1991. Lista de aves. In D.H. Janzen, ed. *Historia Natural de Costa Rica*. M. Chavarría (trad.) 1ra. edición en Español. San José, Costa Rica: Editorial de la Universidad de Costa Rica.

Stiles, F.G. and Skutch, A.F., 1989. *A guide to the birds of Costa Rica*. Ithaca, NY: Cornell University Press.

7 Bio-innovations toward sustainable agriculture

Success stories from India

Nidhi P. Chanana and Neetika W. Chhabra

Introduction

The farming community in India is a marginalized lot, whether they live to the foothills or higher mountains or the eastern flood-prone plains. Owing to poor returns, weather extremities and uncertainties, they have been on the brink of decent survival for decades now. The public instruments of agricultural research and development have taken giant strides in the laboratory, but their arms of extension have not performed as expected. However, there have been some rays of hope in various parts of the country where farmers have broken their shackles of misery and attained self-sustainability. We present here four case-studies from three different regions of India where the Energy and Resources Institute (TERI) has been working for almost a decade now to augment incomes of farming communities through various approaches.

Case study 1: livelihood enhancement in Uttarakhand State, India

As the farmers were looking for respite from these tribulations, they were keen to adopt new crops/technologies that could minimize risks emanating from weather extremities. As the farmers had a limited crop profile, crop diversification to enhance the crop portfolio of the farmers was the first choice. To get started, resource mapping of the farmers was carried out to diagnose deficiencies in soil health of farmlands. A soil testing lab was set up at TERI's research station at TRISHA (TERI's Research Initiative for Sustainable Himalayan Advancement) to carry out routine soil testing procedures for this purpose. A soil health card was provided to the farmers which informed them about the nutrient status of their soils along with recommendations on the dosage of chemical and organic fertilizers for specific crops. Also, a baseline survey was also undertaken to understand the nature of assets of the farmers such as education, family size, size of landholdings, water storage capacity, access to water and road, income levels, infrastructure availability, problems faced by them and the major and minor crops in cultivation. The local population lives in houses which are surrounded by home gardens where crops are grown for self-consumption. The landholdings that they possess are quite small (average size less than an acre) which are basically appendages of their home gardens. Meetings were organized in the villages to introduce the project to the local population based on the interest of the farmers and technical criteria. Looking at the size of the landholdings and the climatic conditions, it was thought to introduce such short duration crops which could fetch a good market price as the growth window in a year was short.

Based on these criteria, medicinal and aromatic crops as well as spice crops were introduced in the villages. The crops that were initiated into farming were scented geranium, parsley, oregano, peppermint, rosemary, turmeric and garlic that were grown in homestead gardens and extensions. Good quality planting material was provided to farmers. Initially, demonstrations were planned on 100 m^2 plots with selected beneficiaries who chose patches of underutilized land for cultivation. These crops were introduced as these could be grown on fallow land and provided crop cover almost round the year, and they required less water and agricultural inputs. In addition, animals did not damage these crops, hence providing better income per unit area from small and fragmented land holdings.

Training was provided to farmers for scientific cultivation methods, harvesting and other operations related to these crops. Moreover, farmers were trained in post-harvest processing of these crops as it led to value addition, increasing their net return. Since these crops were new to the farmers and many of them did not sell in the local wholesale market, it was decided to provide market linkages to the farmers by arranging buyback of the produce. Prices for the whole year were fixed at the time of crop plantation in the field as the risk bearing capacity of the farmers was less. With these initiatives, TERI commenced to develop the aromatic and spice crop value chain. Gradually, farmers started cultivating these cops successfully and even making small net gains. As these aromatic crops contained essential oils, two oil distillation units were established in Supi and Satbunga villages for extraction of essential oil from scented geranium and other crops. Farmers were trained on this process and have been bringing the fresh herbage of scented geranium from their fields to extract the oil for almost seven years. Small biomass-based gasifiers were attached to these units for increasing their sustainability and efficiency. This not just helped in reducing the fuel wood requirement of the units but also improved the quality of the essential oil distilled in the process. This is because when the fuel temperature is regulated, there is no loss of important volatile components and hence the oil quality is enhanced.

As they began to get good returns from these crops, more and more farmers started getting involved in this initiative. As this work was labor intensive and meticulous, the number of women that got involved started increasing. Some of the crops such as garlic, rosemary, parsley and scented geranium were adopted by the farmers as they provided almost INR 3,000 per *nali* (the local land measurement unit, equivalent to 200 square meters).

As the work progressed, it was observed that long periods of dry weather were experienced some years which took a toll on the crop growth and eventually production. Hence, it was thought to provide farmers with innovative and low-cost solutions to address problems related to water dearth in order to strengthen the entire value chain. Various options of low-cost water harnessing and storage were envisaged; the first such solution was polyethylene sheet-lined tanks and rooftop precipitation harvesting infrastructure. Nainital district is known for erratic but heavy rainfall. It was observed that there was no history of organized harvesting and collection of various types of precipitation in the district for farmers. Though concrete/low cost polyethylene sheet-lined tanks had been provided to certain farmers through government initiatives, the numbers were very few and it was the first time when these tanks were connected to rooftop harvesting systems. Therefore, providing rooftop precipitation harvesting infrastructure linked to low-cost tanks would be highly relevant and would be able to meet the water requirement of the crops grown by farmers. Artificial recharge by rooftop harvesting offers an option to store water during the wet months and use when water resources are scarce. Farmers could store 12,500 L of rainwater in these tanks for their use.

Another innovation tested in that area for the first time was the use of drum kit–type drip irrigation technology. This low-cost drip irrigation technology minimizes the initial capital cost at the expense of available cheap labor. The concept is very promising and has proved its potential for poverty alleviation in many areas of the world (Postel et al., 2001). It operates by water gravity from a tank placed 1–1.5 m high. It is a closed-pipe gravity system, a localized method and a seasonal installation for growing vegetables and other horticultural crops on flat or minor slope land. The pressure of the system is very low (0.1–0.2 bar). No external power or electricity is needed. Hence, this technology has the potential to provide critical irrigation to the aromatic crops at various intervals of time. In this manner, this technology provided critical irrigation to the aromatic crops and increased the water use efficiency by almost four times.

These water management solutions have solved the problem of seasonal requisition of water. Farmers could achieve better production by using the additional water harvested which would have otherwise surged down the slopes of the mountains and caused substantial soil erosion. Hence, these devices also contributed to soil conservation measures and led to environment protection.

These crops were cultivated organically using vermicompost and mycorrhizal biofertilizers. Uttarakhand is traditionally an organic state as farmers do not use chemical fertilizers in their terrace farms. Instead they use compost made from oak leaves (the most prominent component of local flora) and cow dung. The women collect oak leaves from the forest in the winters and spread them in the cow shed where the livestock cover it with their waste matter. Here it decomposes slowly and turns into compost. However, because oak leaves are quite hardy, they take a long time to decompose, and many times farmers use partly decomposed matter in their fields as a source of nutrients. TERI established a vermicompost unit utilizing the locally accessible biological waste such as fruit and vegetable waste, cow dung, weeds, oak leaves and poultry waste, and it proved to be a valuable method to generate organic inputs for agriculture. The duration for vermicomposting was found to be shorter when oak leaves were used as substrate. The compost produced was a black, granular, lightweight and humus-rich crumbly powder. The quality of compost, when poultry waste was used as substrate, was better in terms of nutritive value as well as in terms of percent recovery of compost than when field and weed waste acted as a substrate. In case of field and weed waste, there was less decomposition in the final product, which might be due to differences in the cellulose content of the substrate. Farmers were also given training in vermicompost production and were supported to set up their own units in their homestead gardens where they used domestic green waste, field waste and dung to produce vermicompost through earthworms. This is not only sustainable environmentally but even provides a means of income generation. These nutrients promoted the growth of the herbs and also improve the nutritional status of the soil. Efforts are underway to obtain organic certification of these crops so that farmers get a better price realization from these crops. With this vision, TERI facilitated farmers to establish a self-help group (SHG) called P.E.O.P.L.E. (Promotion of Essential Oil Production for Livelihood Enhancement), which is responsible for developing value added products from medicinal and aromatic plants such as dried herbs, essential oils and hydrosols.

Garlic was also one of the spice crops promoted for cultivation in the region. Planting material of the variety Agrifound Parvati with large bulbs and 8–10 cloves per bulb was provided to the farmers. This variety gave good yields under hilly conditions. From a mere 50 kg production in 2008, the production took a gigantic leap to attain the production of 15 tonnes in 2014. Products are being sold under the umbrella brand of *Supi Sugandh* ("The fragrance of Supi").

Signals of transformation

Employment generation

The initiative has heralded new employment opportunities for local youth as they became acquainted with new farming systems and water management technologies to go beyond their conventional systems. They were introduced to micro-entrepreneurship to earn better livelihoods for their families.

Standard of living

As women were involved in most of the operations of this value chain, they had an opportunity to participate and gain knowledge about new activities. This in turn boosted their confidence and even helped reduce their drudgery. It is also steadily leading to women's empowerment and creating a sense of ownership and independence to make decisions.

Income of the rural people

The increase in herbage and subsequently essential oil through efficient water management led to better returns. The farmers attained additional remuneration from the cultivation of the aromatic herbs. On average, the farmers received INR 75,000–200,000 per hectare (ha) annually from cultivation of parsley, oregano, rosemary, garlic and geranium cultivation. Besides, cultivation of aromatic crops assures a well-distributed income throughout the year in comparison to conventional crops.

Environment conservation

The initiative led to conservation of environment through efficient use of water. Water use efficiency increased four times by the use of drum-type drip irrigation. A total of 100,000 L to 150,000 L of rainwater was harvested annually using a rooftop water harvesting system per household, depending whether the roof is sloping or flat. Also, the water saving devices will have a positive effect on the environment as they will lead to efficient natural resource utilization and reduction in wastage of precious water reserves.

TERI's interventions brought fallow/underutilized land under cultivation and water saving by 50% through cultivation of aromatic herbs vis-à-vis that of potato. Lesser use of chemical and fertilizer and use of bio-inputs such as vermicomposting helped enhance soil productivity. Around 10 ha of fallow/underutilized land extension of home gardens was brought under cultivation through TERI's interventions. Presently, crop cover exists in the fields as well as the home gardens of the farmers for the major part of the year as compared to half a year in the conventional scenario. The medicinal and aromatic plants that are being introduced in the homestead gardens have a number of health benefits, especially against respiratory problems. In addition, these, being natural, do not have any side effects. The gasification technology has also led to saving of fuel wood, a high valuable non-renewable energy source as it takes a long time for newly planted trees to reach the harvesting stage and short rotation species are not feasible in this region. With these initiatives, TERI has touched the lives of around 5,000 households spread across many villages of the surrounding area.

Conclusion

Our vision for this initiative has been to develop a self-sustainable system for the farmers which can perform against challenges and can assure better returns. The concept is easily replicable because of the low costs involved. Hence, this approach can be scaled up so that more and more marginal farmers can participate and can gain benefits from this initiative.

Case study 2: promoting traditional crops for food and nutritional security in Uttarakhand State, India

Another initiative that is being executed in some of the villages is the promotion of traditional nutritious crops of the region. During the long period of interaction with the local population, it was learned that certain crops and agronomic practices that had been practiced in homestead gardens from time immemorial had been discontinued for some decades now. The main objective of the proposed project was to revitalize the traditional crops and practices to improve local food security in a sustainable manner.

There are local crops that have been cultivated in homestead gardens of Uttarakhand since ages as it is one of the important centers of crop diversity due to high ecological heterogeneity and high local socio-cultural integration. These crops had been grown for decades or longer but were gradually replaced by cash crops. According to a report by the gene campaign conducted in hilly regions of Uttarakhand (Apetrei, 2012):

> In general, there is a total reduction of the areas and quantities of millets grown. As a type of food becomes less available, social references seem to be immediately adjusted, making it even less likely that the forgotten crop would be later on reintroduced. This means that a cultivated species can be wiped out in only a few generations if, for various reasons, it ceases to be interesting as a crop for a certain society.

Besides millets, pulses cultivation has been limited to subsistence farming. The current crop profile is not sufficient for the daily nutritional requirement of a family. These crops include various cereals such as finger millet, barnyard millet, hog millet and buckwheat; and pulses such as kidney bean, black soybean, adzuki bean and horse gram. These crops are rich sources of many important nutrients and have the added advantages of disease resistance, drought tolerance and other important attributes. Hence, these home gardens were a source of high nutrition for the entire family and contributed to nutritional security of the region. Due to remote location and self-use, farmers have been responsible for maintaining these landraces. However, these very factors have contributed to low yields as well, thus forcing the farmers to switch over to lucrative cash crops. As a result, these landraces have started disappearing. It was becoming imperative to make efforts to conserve these accessions and also document them so as to make them available for future use. Therefore, landraces of these crops were collected and their cultivation was facilitated. This led to the development of a live seed bank for these valuable assets.

From ancient times, farmers of the region cultivated and harvested *Barahanaja* (a mix of twelve grains and pulses, sown and harvested simultaneously in one field) in the monsoon season. This tradition of mixed cropping provided protection against total crop failure and effective instrument of food security. This eco-friendly agronomic system was more sustainable and viable. Some of these crops were amaranth (*Amaranthus oleracea*), buckwheat (*Fagopyrum esculentum*), naked barley (*Hordeum himalayens*), maize (*Zea mays*), kidney bean (*Phaseolus vulgaris*), horse gram (*Macrotyloma uniflorum*), various types of traditional soybean (*Glysine soja, Glysine max, Glysine*), adzuki bean (*Vigna angularis*), black gram (*Vigna mungo*), cowpea (*Vigna unguiculata*), pigeon pea (*Cajanus cajan*), perilla (*Perilla frutescens*), sesame (*Sesamum indicum*), tickweed (*Cleome viscosa*), hemp (*Cannabis sativa*), roselle (*Roselle, Hibiscus subdarifa*) and cucumber (*Cucumis sativus*).

All the crops grown in this system are selected in such a manner to sustain (1) proper germination of seeds from various crops; (2) soil fertility level; (3) production levels during drought or excessive rainfall; and (4) food or feed for the household or its animals. In addition, this cultivation method exploits soil nutrients at different depths and helps maintain insect pests under threshold levels.

However, the yields under such cropping systems have been low and have dwindled further due to low seed replacement rates. The average grain yield of different traditional crops in the region was reported at 10.25 quintals per ha for wheat; 11 quintals per ha for barley; 26.18 quintals per ha for the mixed crop of paddy, barnyard millet and foxtail millet; 18.16 quintals per ha for the mixed crop of finger millet and horse gram; 18.46 quintals per ha for amaranth, and total food grains to be at 18.84 quintals per ha (Semwal et al., 2001; Whittaker, 1988). With the advent of cash crops, these cropping systems started losing their lucrativeness. Efforts were made to revive them but with scientific methodology so that they could help preserve heritage and tradition and yet provide benefits to the growers.

As communication technologies and media become more accessible and urban areas expand, as low-income countries advance to a higher income level, and populations drift toward modern lifestyles and urbane habits, traditional foods are often abandoned. All these factors contribute to degeneration of health status and the overall productivity of people. The worst affected factions are children and women. This group is mainly plagued by malnutrition, low body mass index and micronutrient deficiencies. Assessment surveys about information on existing status of micronutrient deficiencies in the target area helped in the creation of a knowledge database that would cater to the present as well as prospective programs to address these issues. Similarly, assessment of food habits and sociocultural practices facilitated the designing and formulation of strategies for operational support. Promotion of underutilized vegetable crops that are rich in beta-carotene and iron (such as amaranth greens, taro tubers, *kundru* or ivy gourd, *kandali saag*, rhododendrons, *lingoda, kankora* or spiny gourd) through their use in homestead and community gardens helped restore these crops and increased their consumption.

Strategies involving various stakeholders were developed for promotion of purchasing, growing and consuming vegetable and fruit sources rich in micronutrients. Communities, especially women and children, were involved in this activity. They are described later. A number of meetings were organized by TERI with various women SHGs in different villages to create awareness on traditional foods that have very high nutritional value. The shifting pattern of traditional crops toward the present conventional cropping cycle and their effect on human health was also discussed. All the women agreed with the changes in the food system; they showed their interest to incorporate the traditional foods back in their present food habits and requested provision of traditional seed for cultivation in their fields. In the discussion, they shared the experience of their childhood or their parents when they were fed special traditional dishes to overcome minor and serious ailments. Hence, the young women and children benefited from this discussion and their awareness about traditional food and nutritious foods increased.

In another initiative, awareness campaign was started for schools in the area focusing on the importance and nutritive value of traditional crops in Uttarakhand. Quiz competitions were also organized on the same aspects in three schools in which we covered 431 students and awards were given to the winners. In the last phase, a cookery competition was conducted in schools. For this competition, recipes were invited from students and selected recipes were called for the competition.

A survey was carried out to assess the impact of the project interventions with 100 direct beneficiaries of the project, and the following are the salient points that emerged from the survey:

- There was an increase in the number of farmers cultivating kidney bean, horse gram and black soybean by 21%, 56% and 51%, respectively
- Around 97% of the respondents said that their dependency on buying groceries was reduced
- Almost all the respondents were aware of the names of the dishes made from traditional crops as well as their recipe; however, only 60% were aware of the nutritional benefits
- Around 90% of respondents said that they became aware about the traditional cropping system of *Barahnaja* after the project
- Around 46% of the respondents said that they were presently consuming food prepared from traditional crops twice a week.

Hence, promotion of traditional underutilized crops can serve as a major attempt to ensure food and nutritional security while maintaining the local agro-diversity of a region.

Case study 3: livelihood enhancement in Bilaspur district in the state of Himachal Pradesh, India

Background

Bilaspur is a district in Himachal Pradesh. It has a total area of 1,167 km^2 and a population of 380,000. The region has a hilly terrain. It is located in the Shivalik range of the lower Himalayas. TERI started working in Bilaspur district of Himachal Pradesh in 2007. The average land holding of the farmers of the district is 1.06 ha and most of the cultivation in the district is carried out on contours. Agriculture in the region is primarily rain fed and the area receives an average rainfall of 1,106.28 mm. The district has a wheat-maize cropping system. The region is also facing an acute problem of wild animals such as monkeys damaging the crops. There is a dearth of quality planting material of the crops that they were cultivating. The farmland is an extension of the homestead garden of the farmer and the entire family practices agriculture. Few of the crops that are grown by the farmers are for their domestic consumption, such as spice crops, while the rest of the crops are for domestic as well as commercial purposes. The majority of farmers do not find practicing agriculture remunerative, hence they are drifting to other professions. However, if the farmers of the region redesign the farmland according to their homestead garden crops, such as spice crops, farmers would get more income per unit of the land.

TERI's initiatives and their impacts

TERI started its activity in the region in 2007. A survey was conducted to understand the agricultural practices followed by the farmers, the challenges faced by them, their resource evaluation and so forth. Based on the study, TERI proposed a crop profile and set of interventions, suitable for the region, under a project titled Empowerment of Agrarian Population through Demonstration and Evaluation of High Value Plantation Crops, sponsored by the Department of Biotechnology (DBT), Government of India. Under the project, TERI proposed to (1) carry out soil health analysis at farmers' fields and thereafter provide need-based recommendations; (2) build awareness among the farming communities on the application of biotechnology for enhanced crop productivity; (3) provide high-quality planting material both conventionally raised as well as tissue cultured for cash crops, such as potato; clonally propagated stocks of fruit crops, namely banana and pineapple; spice crops, namely ginger, turmeric and large cardamom, for distribution to farmers of the selected districts; (4) train and build capacity building of farmers on scientific method of cultivation, on-farm production of biofertilizers such as vermicompost, and mycorrhizae, and biopesticides such as *Trichoderma, Pseudomonas,* etc.; and (5) promote agri-based enterprises among the rural population through demonstration and training.

To begin with, a large numbers of farmer meetings were organized to make the farmers aware of how biotechnology can help them in increasing their farm productivity and income. New crops and new varieties of existing crops were introduced to promote crop diversification and help farmers resolve the issue of man–animal conflict. The varieties were collected from all over India. Few of them, namely banana and ginger, were initiated under in vitro conditions and tissue culture–raised plants were produced. Hardening of the plants was done at TERI's DBT-sponsored tissue culture facility at the Micropropagation Technology Park (MTP). Secondary hardening was carried out at low-cost shade houses constructed at Bilaspur district, Himachal Pradesh. Plants were distributed to selected beneficiaries. Training programs were conducted for farmers on the scientific method of cultivation of the selected crops, the use and on-farm production of bio-fertilizers and biopesticides, entrepreneurship development and so forth. Demonstration and evaluation trials were carried out to motivate farmers to adopt these interventions. The following is a detailed account of some of the crops/interventions demonstrated in the region.

Turmeric: Turmeric was grown by the farmers of the region only for domestic consumption. The farmers were cultivating a local variety. Leftover harvest from the previous year served as planting material for the subsequent year. In 2007, TERI introduced ten varieties, collected from all over India: Ambika, Sugandha, DPT-1, DPT-2, T-4, Palam Lalima, Pant Peetabh, Kerala (Alleppey finger), Cuddapah and Megha. These varieties were grown in small demonstration plots and their yields were evaluated. Farmers were provided with both conventional as well as tissue culture-raised planting material. Among all the varieties tested, DPT-2 (47.1 q/acre) and DPT-1 (46.1 q/acre) exhibited the highest response, followed by Ambika (43.8 q/acre) and Pant Peetabh (40.7 q/acre). The response of all the introduced varieties was 30%–90% higher than the local variety (25 q/acre). To increase the yields further, rhizomes were treated with mycorrhiza at the time of sowing. Mycorrhiza had significant effect on yields in few varieties. A comparative yield analysis of conventional vis-à-vis tissue culture raised plants indicated that barring first year, the yields from tissue culture raised plants was higher than that of conventional plants. An intercrop model was also demonstrated to the farmers where pigeon pea was intercropped between turmeric. An increase in income by almost 34%–38% was obtained in the turmeric-pigeon pea intercropping model versus the turmeric monocropping model. An impact assessment over three years (2007–2010) showed that average area per farmer under turmeric cultivation increased by 250% and the number of farmers cultivating turmeric increased by 146%.

Potato: Before TERI's intervention, in Bilaspur, the Kufri Jyoti was the only potato variety cultivated and at a very small scale. TERI carried out a resource evaluation of the farmers' fields and thereafter selected the farmers. New varieties were introduced, including Surya, Himalini, Himsona, Pukhraj, Chipsona, BJ-1, Shelja and others. Planting material of both tissue culture–raised mini-tubers and conventional tubers were provided to the farmers. Introduction of new varieties led to an increase in yield and income by 7%–16%. Yield from tissue cultured, first generation mini tubers (e.g., Chipsona 53.1 q/acre) was comparable with the yield from conventional tubers (56.6 q/acre). However, considerable improvement in yield was obtained from second- and third-generation tubers of tissue culture–raised plants. In fact, the yield was above 50% that of conventional. An impact assessment of three years (2007–2010) showed that the area under potato cultivation increased by 195% and the number of beneficiaries by 44%. To help the farmers grow potato sustainably, TERI helped the farmers establish producer groups. The producer groups collected the seed potatoes from the farmers, stored it in cold storage and made the tubers available to farmers for sowing in the planting season. In addition, these producer groups also helped the fellow farmers in sale of their produce. A total of ten producer groups were established in the region.

Low-cost vermicomposting units: In the Bilaspur district, vermicomposting was being done as small-scale, backyard composting using kitchen waste as a raw material. However, the percentage of farmers practicing vermicomposting was very small. TERI demonstrated the construction of low-cost units. Three units having 52 pits of 5 × 4 × 2 ft were constructed. Since the terrain where the units were constructed was such that there was no problem of waterlogging, the pits were dug in soil and filled with different feed stocks. The units were covered with a thatched roof to protect worms from direct sunlight. Different locally available feed stocks were used: agro-wastes, ageratum, bamboo leaves, dry grass, farmyard manure (pure), kitchen wastes and lantana. Earthworms, *Eisenia foetida*, were added to the feedstock. Vermicompost made from these feed stocks was tested on maize. It was observed that among the different feed stocks tested, vermicompost from lantana showed the highest yield response, followed by that of kitchen waste. The results encouraged many farmers to take up vermicomposting in their backyards as a zero-investment, crop-friendly intervention.

Case study 4: livelihood enhancement in Kamrup, Morigaon and Dhubri districts in the state of Assam, India

Background

The state of Assam is endowed with rich natural resources, however its economy represents poverty among abundance. It lags behind the rest of the country in several aspects. Agriculture is the main occupation of the region, providing employment and livelihood to more than 50% of the workforce. However, according to the Economic Survey of Assam conducted in 2014–2015, the percent contribution of agriculture and allied activities to the gross state domestic product (GSDP) has declined steadily (Directorate of Economics and Statistics, 2017). There are many reasons for this trend and it affects the overall growth both directly as well as indirectly. One pertinent issue affecting growth and development is that the total agricultural land of the state is 2.81 million ha but the average landholding is only 1.10 ha. As such, more than 85% of the farmers are small and marginal, with an average landholding of 0.63 ha, which is not consolidated but scattered. Every household in Assam has a homestead garden, where the farmers grow all the local spices which can be grown in the region: turmeric, ginger, black pepper, hot pepper and so forth. In addition to this, the farmers also grow arecanut, coconut, banana and jackfruit in their gardens. In the main farming land, rice is the main cultivated crop. The other prominent crops grown in the region are arecanut followed by coconut. Most of the region is rain fed. The climate of the region is well-suited for paddy cultivation, which occupied more than 92% of the cultivated area under food grains. In the region, farmers do not have access to new agro-biotechnologies and quality planting material. Most of the villages are located in the interiors, where there is no decent public transportation. Farmers in Assam practice rain-fed cultivation and the available underground water is not of good quality. Assam also experiences extremes of weather conditions. There would be incessant rains leading to floods as well as dry spells. This badly disturbs their agricultural and personal water requirements. Crop production is negatively affected under both weather conditions. The region has lot of climatic and law and order challenges. Owing to this, there is not much development in the area. As most of the villages are located in the interiors, they do not have access to information and technical know-how on agri-technologies.

TERI's initiatives and their impacts

TERI has a centre in Assam and is working on various aspects including agri-extension. The author started working in the region in 2007. A survey of the area revealed inadequate utilization of the agricultural land, farmers practicing sustenance farming, lack of quality planting material and so forth. The work presented in the chapter was done under two projects sponsored by DBT, Government of India, from 2007 to 2015: (1) Empowerment of Agrarian Population through Demonstration and Evaluation of High Value Plantation Crops and (2) Agri-Biotechnologies in Livelihood Enhancement of Agrarian Population in Kamrup, Morigaon and Dhubri Districts of Assam. The projects were carried out in two phases. In the first, elite phase, tissue-cultured varieties of banana, pineapple, turmeric and ginger were tested in demonstration and evaluation trials, in only the Kamrup district of Assam. Farmers were provided with high-quality planting material after carrying out resource evaluation of the farmer's field. Training on their scientific methods of cultivation, pest management, intercropping, on-farm production and application of bio-fertilizers, biopesticides was carried out. Although the area faces acute weather condition of excessive rainfall leading to heavy floods and intermittent dry spells, certain crops, namely banana, ginger and turmeric, were well received by the farmers. Owing to the success of the first phase, in the second phase, the area was extended to additional districts, Morigaon and Dhubri. In the second phase, more emphasis was given to strengthening the turmeric value chain by setting up of turmeric curing and dehydration units and on waste management by establishing vermicomposting units and a banana fiber extracting unit. The following is a detailed account of few of the successful interventions.

Turmeric: In Assam, most of the farmers grew turmeric in their homestead gardens for their domestic consumption. TERI encouraged the farmers to cultivate turmeric at a larger scale either alone or in SHGs. Farmers were encouraged to grow turmeric on their fallow or underutilized land. Quality planting material of the Lakadong variety of turmeric, which has high curcumin content, was provided to the farmers. In the first phase of the activities, TERI focused on only one district, Kamrup, where the farmers were provided with the planting material and were given training on scientific method of cultivation. To have more yields per unit of land, some of the farmers grew turmeric as an intercrop with arecanut. Since turmeric is a shade-loving plant, it grew well as an intercrop. The yields were comparable to plants growing in open areas. A total of 220 beneficiaries were provided with quality planting material. The farmers sold their produce in the local market. However, it was felt that there was a need to strengthen the value chain and that instead of raw turmeric, processed turmeric would fetch a better price. Keeping that in view and the need to spread the awareness of renumerative potential of the crop, in the next phase TERI extended the work in three districts: the districts of Kamrup, Dhubri and Morigaon in Assam. Farmers were provided with high-quality planting material of 'Lakadong' turmeric. Training was imparted to farmers on the scientific method of cultivation. A comparative analysis of yield was carried out in all the three districts. The yield varied in all the three districts. The highest yields were obtained in Morigaon district at 5,012.4 kg/acre, followed by Dhubri at 4,043 kg/acre and Kamrup at 2,537 kg/acre; the average yield exceeded the district's average.

A turmeric processing unit was also set up to help farmers process their produce and sell it on a higher price. The turmeric processing unit has a washer cum polisher, a boiler and a grinder. For processing, fresh turmeric rhizomes are collected. They are washed in washer, removing the soil and dead roots. The rhizomes are then loaded in a mesh bucket and autoclaved for 15 minutes. Thereafter, the rhizomes were sun-dried till they were completely dehydrated. The processed turmeric gives a metallic sound after breaking. The rhizomes were then loaded in a polisher, where they were polished and the dead skin and roots were sloughed off. Thereafter, the rhizomes were ground in a mill and ready to use as a turmeric powder.

Training was imparted to the farmers, who processed their raw turmeric in the unit. A few of the farmers sold their produce in the local market and the rest was bought back by TERI and sold in retail.

Banana: The most popular varieties in the Assam region are Malbhog and Chinichampa. However, most of the plantations were devastated because of panama wilt disease. The disease is gradually spreading to the virgin area because farmers are using suckers from trees growing in infected plantations. There was not a single source in the region which was providing farmers with the clean, quality planting material. During TERI's activities in the region, farmers were provided with tissue culture–raised plants of the Grand Naine variety of banana. The activity faced many problems owing to the weather conditions – flooding in some area and lack of water in others. The water availability in certain regions worsened to such an extent that people were not having adequate water to carry out even their routine morning ablutions. The selection of the beneficiaries for growing banana in fact could not have been felt more important than during this period. However, the local weather conditions were so dynamic that even selection did not prove fruitful, as in the subsequent year the same areas were flooded that were initially facing water-problem or did not face the problem of flooding earlier. Tissue culture–raised plants were provided to farmers in the districts of Kamrup, Morigaon and Dhubri. Farmers were motivated to undertake plantations in the fallow land or as an intercrop between arecanut plantations, and training was provided on cultural practices on banana. A total of 221 beneficiaries grew tissue culture–raised banana and a total of 66.7 acres was covered. Through banana cultivations farmers earned an average revenue of INR 94,900 per acre.

Utilizing waste: In order to supplement the income of farmers of the Kamrup district of Assam, we initiated an activity of extraction of fibers from spent banana stems. In banana, after flowering and fruiting, the spent stem is of no use. The next year fruiting is initiated from the daughter suckers which grow up after the main stem has borne fruit. The spent stem is cut off and is a source of pollution. During our project activities we realized that we can help the farmers utilize these spent stems.

Banana fibers have wide use in handicrafts, dress material, currency notes, paper and so forth. The fibers can be extracted chemically, manually or mechanically. The fibers extracted manually are the best quality fiber, followed by mechanical extraction. However, manual extraction is laborious and time-consuming. Thus TERI, with the help of DBT, established a banana fiber extraction unit. The unit was easy to use. The spent stem was cut in thin strips, and each strip was passed through the machine twice or thrice. The machine had a blade fitted at an angle. All the plant tissues were separated from the fibers, and the extracted fibers were then left to dry. After drying, these fibers were then used for a variety of purposes. The farmers were trained on how to use the machine and develop products from the extracted fibers. This is still a new technology in the region, but it is gaining popularity among farmers, and in the near future it is expected that more farmers will adopt this technology.

Conclusion

The farmers of the region were suffering immensely because of their abysmal economic conditions, illiteracy, non-exposure to the modern agri-interventions, lack of end-to-end linkages and so forth. The adoption of region-specific approaches led to diversification of the existing crop profiles, more produce per unit of land, better land soil and water use efficiency, and entrepreneurship. Most of the land that was used was either fallow or underutilized. As each household in the region has a homestead garden, the fallow and underutilized lands in the homestead gardens were cultivated systematically to give commercial yields. This, in turn, translated into livelihood enhancement of the agrarian population of all the project sites.

References

Apetrei, C., 2012. *Food security and millet cultivation in the Kumaon region of Uttarakhand research report for gene campaign.* Gene Campaign, New Delhi, India.

Directorate of Economics and Statistics, 2017. *Economic survey, Assam (2014–15).* Directorate of Economics and Statistics, Govt. of Assam, 2017. [Online] Available at: https://des. assam.gov.in/documents-detail/economic-survey [Accessed 30 November 2019].

Postel, S., Polak, P., Gonzales, F. and Keller, J., 2001. Drip irrigation for small farmers: A new initiative to alleviate hunger and poverty. *Water International,* 26(1), pp. 3–13.

Semwal, R.L., Maikuri, R.K. and Rao, K.S., 2001. Agriculture, ecology, practices and productivity. In O.P. Kandari and O.P. Gusain, eds. *Garhwal Himalaya: Nature culture and society.* Srinagar, Garhwal: Transmedia Publication, pp. 259–275.

Whittaker, W., 1988. Migration and agrarian change in Garhwal district, U.P. In T.P. Bayliss-Smith and S. Wanmali, eds. *Understanding green revolutions: Agrarian change and development planning in South Asia.* Cambridge: Cambridge University Press, pp. 151–165.

8 Home gardens as a resilience strategy for enhancing food security and livelihoods in post-crisis situations

A case study of Sri Lanka

D. Hashini Galhena Dissanayake, Gunasingham Mikunthan and Linda Racioppi

In many parts of the world, violent conflicts and disasters have had devastating effects on agriculture and, by extension, on food security and livelihood options. Such crises too often leave economies, particularly agricultural sectors, in shambles: access to agricultural inputs shrivels, agricultural productivity drops, infrastructure is destroyed and rural livelihoods are ruined. In such circumstances, finding post-crisis agricultural strategies that increase food sovereignty is critically important. This chapter is an overview of the literature on food security and food sovereignty and their relationship to livelihoods as well as on crises. The discussion then proceeds to showcase home gardens as a resilience strategy within the local food systems and in the broader context of agricultural development in countries that have suffered from various forms of crisis and disaster (Racioppi and Rajagopalan, 2016). The chapter concludes by examining the case of Sri Lanka, a country that has emerged from a devastating civil war, has experienced natural disasters and has suffered from food insecurity, especially in those areas most affected by conflict and natural disasters. The experiences with home gardens as a strategy to augment agricultural resilience and encourage food sovereignty for poor households from Sri Lanka suggests that home gardens can provide for household food needs in the face of serious food insecurity and can also contribute to the improvement of rural livelihoods.

Food security and food sovereignty

Since its inception in the early 1970s emerging from global concerns for available food stocks, the concept of food security has evolved to include numerous interpretations. The earliest reference to the concept of food security was noted at the 1975 World Food Conference.[1] This was followed by a series of debates assessing and adapting the core principles highlighted in the original conceptualization to fit a range of circumstances. By 1993, Smith et al. compiled an annotated list of around 200 such characterizations (Smith et al., 1993). While some ideas may be context-specific, most broadly include issues related to food availability, access, utilization and more recently, sustainability. Although it has undergone multiple revisions, the food security definition commonly adopted today by the Food and Agriculture Organization of the United Nations (FAO) and other international development organizations originally stemmed from the Rome Declaration of World Food Security in 1996 – a situation when "all people, at all times, have physical, social and economic access to sufficient, safe and nutritious food that meets their dietary needs and food preferences for an active and healthy life" (FAO, 2001).

A more recent but closely related distinction is food sovereignty, a concept that developed in response to concerns about the tendency of the food security approach to overlook the importance of local food systems in addressing issues of hunger and malnourishment. First highlighted at the 1996 World Food Summit by a civic organization called Vía Campesina, food sovereignty is often seen as "the right of each nation to maintain and develop its own capacity to produce its basic foods respecting cultural and productive diversity" (La Vía Campesina, 1996). The literature on food sovereignty claims that it is a precondition for food security. One of the complaints by food sovereignty proponents is that food security does not specify the source of food if the food is available, accessible and absorbable by those in need on a sustainable basis. Thus, in situations where the cost of producing food is high, a nation could opt to import cheap food from world markets or depend on food aid to be food secure. While this may be the preferred option in the short run when the country lags in its agricultural production capacity, countries are highly susceptible to global food stock and price fluctuations. Moreover, in most developing countries that are characterized by a large smallholder farming population, cheap imported foods can negatively impact local producers as they may not be in a position to compete. In comparison, the food sovereignty viewpoint emphasizes the importance of building and maintaining domestic food systems to meet local demand, and it calls for political and social reforms such as equitable distribution of resources.

Both the food security and food sovereignty perspectives have their place in addressing food insecurity wrought by crisis and disaster situations. While considerations of access, availability and utilization (food security) are clearly important for those living in war- or disaster-affected areas, it is also the case that recovery from crises often requires the reliance on and revitalization of local food systems to meet the immediate needs. That is, as local food production plunges during a crisis, communities become vulnerable to global food shocks and their path to food sovereignty is hampered. At the same time, after a crisis, agricultural renewal may also afford rural households the opportunity to enhance livelihoods and reduce food insecurity.

Livelihoods

Chambers and Conway state that a livelihood entails "people, their capabilities and their means of living, including food, income and assets" (Chambers and

Conway, 1991, p. 1). A more recent conceptualization expands Chambers and Conway's definition to view livelihoods as a system consisting of inputs, purpose, activities, agency, quality (vulnerability or sustainability), environment/ context and location, with the output being the livelihood (Niehof and Price, 2001). A livelihood becomes sustainable when an individual, or more commonly a household, is able to manage stresses and shocks and continue to enhance capacities and capabilities (Niehof, 2004).

Food security and livelihoods have a complex but reciprocal relationship. Where there is food insecurity there is generally disruption to livelihoods, as hunger and malnutrition result in poor mental and physical growth and susceptibility to illness that impair human productivity as well as engagement in the labor force. Likewise, when livelihoods are erratic, whether on the farm or elsewhere, families are at risk of food insecurity due to inadequate food availability but more importantly due to limited access and utilization of available food caused by poverty.

Unemployment and the resulting poverty have serious implications on food sovereignty. More than 70% of the rural population in developing countries falls below the poverty line and consists mostly of resource-poor farmers or farm laborers (International Fund for Agricultural Development, 2010; Olinto et al., 2013). Various barriers to agricultural production such as access to quality inputs, resource limitations, cheap food imports and so forth limit livelihood potential, further aggravate poverty and create conditions for food insecurity.

Not surprisingly, livelihoods, as the systems perspective suggests, are complex. Depending on social, political, economic and environmental contexts, livelihood capacities of the poor often decline in the face of natural and human-made crises. Even under non-crisis situations, women, lower caste and tribal communities are found to be the most food insecure and poverty stricken due to social and economic barriers (Mukherjee et al., 2010). The livelihood capacities of women-headed households as well as comprising marginalized peoples are typically the most affected by natural or human-made disasters because they are the most vulnerable and often have greatest difficulty in accessing inputs necessary to meet their production needs (Pehu et al., 2009). However, even the livelihoods of those who are not usually economically or socially disadvantaged can be at risk in the face of crisis, as vulnerability rises for all as the environmental or sociopolitical context deteriorates.

Crises

While the tangled challenges of hunger and poverty can stem from multiple causes, crises of three types can be particularly problematic for food insecurity: economic crises, natural disasters and socio-political crises. There is a large body of literature that explores the economic causes of global hunger and poverty. Some of that scholarship focuses on the global financial crisis of 2008 and demonstrates how it impacted the food and livelihood security of an array of stakeholders across the agricultural value chain (Brinkman et al., 2010). The magnitude of the effects was particularly severe for developing countries that had higher dependence on export markets and foreign investments for capital creation. The agriculture sector specifically suffered immensely as the price of inputs doubled and farmers were unable to benefit from rising commodity prices. As a consequence, in 2009 another 100 million people became food insecure (FAO, 2009).

A growing scholarship has also emerged to examine the role of natural hazards and disasters on global agriculture, rural societies and livelihoods and food security. For instance, many places are at high risk for earthquake, flooding, drought or other natural hazards that affect food production and consumption. In some instances, the inhabitants of an area that experiences annual flooding may have adapted agricultural systems to cope with this risk. However, massive flooding can create crisis situations that go well beyond ordinary coping mechanisms, destroying crops, wiping out homes and causing human and animal casualties. According to Comfort, Boin and Demchak, when such risks rise to crisis level, we can speak of disasters – "low chance, high impact events: [that present] urgent threats to societal core values and life-sustaining systems that typically require governmental intervention under conditions of deep uncertainty" (2010, p. 2). "Natural" disasters, like the 2004 tsunami that devastated Indonesia, India, Sri Lanka and Thailand, or the 2016 Sri Lankan floods and landslides prompt governments, non-governmental organizations (NGOs) and civil society to step in to respond to basic needs for food and shelter and with the Hyogo process to encourage disaster preparedness at the community level. Nonetheless, large-scale events as well as smaller, more regularized hazards can disrupt agricultural economies and food production, not only by wiping out crops for the season but by displacing farmers, eroding and salination of soil and disrupting infrastructure and transportation. Yet, the consequences of such risks for household livelihoods and food security are not necessarily equally distributed, even within a particular hazard-prone area. As a seminal piece by Cannon points up, "social systems . . . generate unequal exposure to risk by making some groups of people, some individuals and some societies more prone to hazards than others" Cannon (1994, p. 14). To make matters worse, farming, livestock and fisheries systems upon which the livelihood of a majority of low-income populations are reliant are rapidly deteriorating due to over-exploitation of natural resources. As might be expected, poor farmers with inferior and risk/disaster-exposed land, inadequate or unreliable water supply and less access to other necessary inputs are least able to cope with such recurrent hazards. Communities in remote areas or areas that lack infrastructure are at risk of not receiving aid in a timely manner or securing assistance from the government. Because these natural phenomena are becoming more prolonged, frequent and unpredictable due to emerging effects of climate change, they are increasingly having devastating outcomes on communities that are dependent on agriculture systems for food and livelihoods (Devereux and Edwards, 2004). As such, poverty-stricken households may be one disaster away from hunger or starvation.

In some countries in Asia and Africa, particularly those with an agricultural economic base, food and income problems are further aggravated by a third set of factors – sociopolitical conflicts. As Collier (2009) has shown, conflict (particularly protracted conflict) is one of the major obstacles to economic development. According to the Armed Conflict Database's most recent figures, in 2015 the world suffered from 40 ongoing conflicts, with 16 of those in Asia (International Institute for Strategic Studies, 2019). Violent conflict and war are particularly pernicious contributors to food insecurity. Whereas the international community is likely to stage a humanitarian intervention and provide food aid as a result of a natural disaster, it is less likely to be effective in doing so during a civil war, even when it is willing to intervene. One of the early, comprehensive reports on the impact of conflict on food security and the agricultural sector published by the Food and Agriculture Organization of the United Nations in 2003 found a wide array of human and environmental consequences that undermine food security: human deaths and injuries, refugees caused by forced migration, acute food shortages and malnutrition, disruption of health care services, the collapse of infrastructure and transportation networks, decline in livestock as well as the disappearance of agricultural extension systems and deterioration of irrigation systems. The report states, "The agricultural sector is usually badly hit in conflicts in developing countries, for lack of inputs and the use of anti-personnel mines which make farmers' agricultural land unusable" (Teodosijević, 2003, p. 10). As a result, war and conflict are acute crises that damage agricultural output, generally reducing it by at least 7%. A United Nations Environmental Program report summarizes the range of consequences: it claims that conflicts and war

> disrupt food production through physical destruction and plundering of crops and livestock, harvests and food reserves; they prevent and discourage farming; they interrupt the lines of transportation through which food exchanges and even humanitarian relief, take place; they destroy farm capital, conscript young and able-bodied males, taking them away from farm work and suppress income earning occupations. The impact of conflicts on food security often lasts long after the violence has subsided, because assets have been destroyed, people killed or maimed, populations displaced, the environment damaged and health, education and social services shattered; still more awesome are the landmines which litter agricultural land, kill and cripple people and deter them from farming for years – even decades – after all violence has ceased.
>
> (World Food Summit Plan of Action 2002)

While such effects are widespread, they are not universal. For example, landmines and infrastructure destruction have been major problems in conflict-affected areas such as Sri Lanka, but their effects have not been universal. Indeed, conflicts that are confined to particular regions of a country (such as war in Sri Lanka) are less likely to breed widespread food insecurity than those where violence is geographically extensive. Still, the consequences for agriculture in war-affected areas are tremendous, and it can take years for these areas to recover. Yet the food and income need of peoples are immediate. Thus, encouraging agricultural practices that meet those needs is critical to their survival and to post-conflict stability.

Figure 8.1 captures a framework that presents interactions between the tri-factors: prolonged hunger, irregular livelihoods and crises or disasters. One or more of these factors can have a direct or inverse impact on each other resulting in harmful consequences. For instance, prolonged hunger can lead to crisis and conversely crisis can exacerbate hunger. The outcome from their

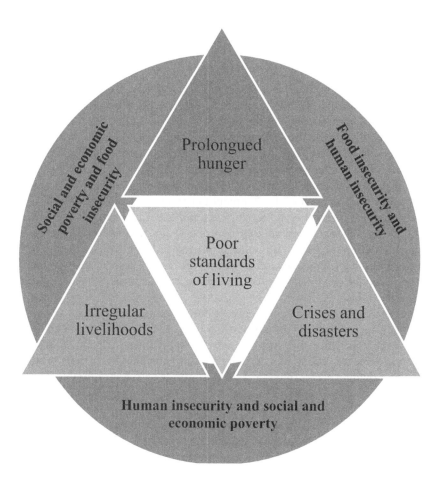

Figure 8.1 Interactions between food insecurity, irregular livelihoods and crisis/disasters

Source: Prepared by authors.

interactions can be equally or more devastating to human wellbeing. However, the intensity of risk will vary depending on the current level of vulnerability and resilience. Frequently, smallholder farmers and the poor are the most vulnerable victims to disasters due to worsened social and economic conditions, destruction of infrastructure, loss of assets or resources for livelihoods, escalation of violence and turmoil and disruption of law and order. Those who experience poverty resulting from erratic livelihoods and food insecurity from chronic or transitory hunger face marginalization, exclusion, and financial constraints (Food and Agriculture Organization of the United Nations, the International Fund for Agricultural Development, the United Nations Children's Fund, the World Food Programme, and or the World Health Organization, 2019). Historical events attest to the numerous occasions when starvation and lack of economic opportunities led to turmoil and unrest as well as disasters leading to hunger and poverty.

Resilience in the face of crisis: the role of home gardens

Under these circumstances, it is not surprising that scholars and practitioners have begun to think in terms of resilience in agriculture and food systems. The term resilience was first used in the field of ecology studies by C. S. Holling in his seminal 1973 article to mark the ability of systems to maintain and to adapt to external conditions), leading them to generate a definition that includes both robustness (maintenance despite external shocks) and adaptability (coping and even improvement in the face of external shocks). These components are reflected in the approach of development practitioners. For instance, the UK's Department for International Development (DFID), like many other development agencies, has adopted an approach that sees resilience as "the ability of countries, communities and households to manage change, by maintaining or transforming living standards in the face of shocks or stresses – such as earthquakes, drought or violent conflict – without compromising their long-term prospects" (DFID, 2011).

As related to agriculture, resilience has most often centered on the notion of making agricultural systems adaptable to climate change. The United States Agency for International Development's Building Climate Change Resilience and Food Security Program focuses on enhancing smallholder farmers' productivity by providing access to best practices as well as new agricultural technologies (United States Agency for International Development (USAID), 2014). Aliou Boly (2013), on the other hand, asserts that this process requires no new technologies but simply applying existing knowledge to ensure "the development of stronger, empowered and resilient communities" (Boly, 2013).

Global experiences suggest that countries emerging from crises remain vulnerable to future turmoil and are trapped in problems resulting from economic collapse, food insecurity, poverty, displacement and other socioeconomic stressors. Thus the processes of rebuilding and reconciling conflict-affected societies and recovering from disasters are colossal tasks and entail tremendous efforts at

Table 8.1 Five aspects of the rehabilitation and development process

Physical	Rehabilitating and reconstructing infrastructure
Economic	Restoring income earning opportunities in agriculture, self-employment in small businesses or access to employment or casual labor
Social	Restoring social ties within the community, households, rebuilding trust and confidence, improving health and education
Political	Offering opportunities for participation and establishing linkages with the politico-administrative system
Environmental	Addressing ecological impacts of war and deterioration of natural resources from overexploitation

Source: Korf and Bauer (2002).

various levels. Immediate interventions must produce quick and visible impacts on the ground to address urgent needs and build confidence in communities. The subsequent strategies should further resolve and mitigate future risk of social and political unrest (Korf and Bauer, 2002). Korf and Bauer (2002) identify five aspects that must be considered in the rehabilitation and development process (Table 8.1).

In order to make rapid impacts on the lives of the conflict-affected people, rebuilding strategies must take into account the infrastructural and resource limitations that exist after war. Although food aid can be a viable immediate strategy until the local infrastructure and production systems are restored, the distribution of such relief may be hampered by lack of institutions and resources to ensure equitable dispersal. Bandarin et al. point out that in post-conflict situations, interventions that have an indigenous and traditional linkage to the target population have proved to be more effective than advanced new practices in reducing hunger and poverty (Bandarin et al., 2011). In a number of post-crisis contexts, home garden projects have been introduced as a food security and livelihood enhancement strategy. Home gardening is an age-old and time-tested subsistence food production structure that predates modern agriculture. Home gardens have been making important contributions to family diet and income throughout the world (Marsh, 1998). In the literature, home gardens are defined as: "a small-scale production system supplying plant and animal consumption and utilitarian items either not obtainable, affordable, or readily available through retail markets, field cultivation, hunting, gathering, fishing and wage earning" (Niñez, 1984).

Odebode (2006) states that home gardening involves the cultivation of a small patch of land near the family home. They can be established in both rural and urban areas. Home gardens are characterized by dense composition of species where a mix of plants such as vegetables, fruits, plantation crops, spices, herbs, ornamental and medicinal plants as well as livestock are raised. The outputs from the garden provide a supplemental source of food, nutrition and income for the household. Typically, they are arranged and tended by the members of the household during their spare time; women, children and elders are the most active managers of these gardens. Home gardens require very few inputs and simple tools. Most families use resources that can be found in around the household or from neighbors. The productivity of the garden is often based on time availability, experience and resource capacity of the household (Mendez et al., 2001). The multiple benefits of home gardening are highlighted in literature. Enhancement of food, nutritional and income security, mitigation of risks and environmental benefits are commonly discussed in various case studies (Landon-Lane, 2011). Galhena et al. (2013) summarized the benefits of home gardening into three categories: social, economic, and environmental benefits.

Home gardens in post-conflict Sri Lanka

Home gardening in Sri Lanka predates the 1970s, when the structure and functions of the Kandyan home gardens were first documented in scientific literature (McConnell and Dharmapala, 1973). It is a common land use practice in the country and has been integrated into daily life. Sri Lankan home gardens typically involve complex annual and perennial cropping and livestock production systems that provide multiple benefits to the household depending on their composition. Mattsson et al. (2018) conducted an extensive survey of literature covering home gardens in Sri Lanka and found several major functions, similar to those noted more generally by Galhena et al. (2013): to provide foods to supplement food security, deliver ecosystem services, act as a safety net when food is in short supply and contribute to caloric intake and nutrition. Thus, home gardens were not only a familiar facet of daily life in much of Sri Lanka, they also functioned as a means to enhance food security which was critical in the wake of the country's disaster in its coastal areas due to the 2004 tsunami and then the ending of the ethnic war in 2009. Both events were notable tragedies that affected the food security and well-being of the people. Over the thirty years of the conflict, thousands were killed, many were disabled and wounded and hundreds of thousands in northern and eastern areas were displaced from their homes. During the most acute phases of the war, many were forced to flee with few belongings to safer locations. And even in the aftermath of the conflict, stresses continued as they were resettled in areas of the northern and eastern provinces.

The survivors returned to their previous dwelling if those areas were rehabilitated; otherwise, they were relocated in areas away from their original home in locations that were cleared of land mines or forests. Some families lost their male heads of household, creating increased burdens on survivors. Family members often had to rely on women and children, some of them traumatized and injured, to provide income for food, housing and healthcare. In the new location, they had to reestablish their lives and livelihoods, however, immediately after the conflict, livelihood options were minimal. The aftermath of the war created a scenario in which people lost their movable and immovable assets and were left to find income generating opportunities that would lead to a sustainable future. Food security was an acute problem. Both availability and access to food were constrained by several factors including the drop in production area, inability to farm due to safety issues (land mines), low food supplies, poverty, poor access to markets and so forth. Often there would be no food or inadequate amounts of food and even when food was available, it was costly. It was difficult for a family to afford three meals a day. Moreover, what they could afford lacked nutritive value. As such, in the period following the end of the conflict, limited food stocks and access problems coupled with issues related to water, sanitation and hygiene had damaging impacts on food and nutritional status of the war-affected people. Unfortunately, some families that were near the coastline in the North East and East endured the devastating effects of the conflict and had earlier suffered from the 2004 tsunami.

Assistance from government, non-governmental organizations and international non-governmental organizations did to some extent support these households to survive in the immediate aftermath of the war; however, the aide they provided was hardly enough to satisfy people's basic requirements. As a result, many families were forced to spend time and energy in a quest for food security (obtaining sufficient amounts of food, searching for nutritious foods for a balanced diet to meet caloric intake needs and stave off disease) and stable livelihoods and income.

Relief through gardening at different levels

As a way of addressing issues highlighted in the previous section, the Sri Lankan government created a broad-based initiative to promote home gardening in post-crisis areas. This initiative targeted three key venues – government offices, schools and homesteads. The Government of Sri Lanka's policy was to maximize agricultural production in the country with the goal of achieving food sovereignty through a multi-tier approach while transforming the country into a "Toxin-Free Nation," with the aim of ultimately switching to organic production. Thus, the programs for office gardens, school gardens and home gardens encouraged growing crops without the use of toxic agrochemicals such as pesticides and fertilizers. The Presidential Secretariat channeled its directives to various ministries that worked to implement the initiative through a range of local, public institutions. The flow diagram (Figure 8.2) depicts organizational mechanism deployed for this task.

The gardening programs at both the government offices and public schools emphasized waste management and recycling methods as well as growing plants in limited spaces. The Department of Agriculture of the Ministry of Agriculture conducted exhibitions countrywide to demonstrate how plants can be grown in small spaces. Through these efforts, the government hoped to encourage people to do the same at home. In order

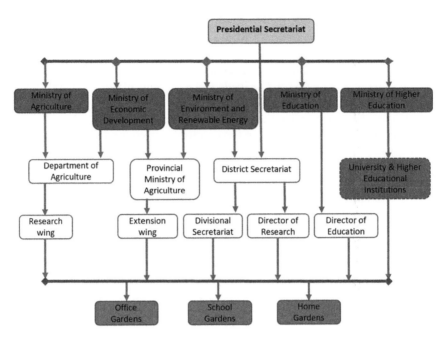

Figure 8.2 Flow diagram showing the different channels and organizations/units involved in promoting home gardening

Source: Prepared by authors.

to support gardening at home, the Department of Agriculture supplied required planting materials. In addition, agricultural instructors and *Grama Niladharis* (village counselors) provided advice on how to grow crops on limited land. In addition, the Sri Lankan military also promoted home gardening programs among the resettled families and conducted competitions to increase enthusiasm.

Case study 1: an exemplary home garden established in Jaffna, Sri Lanka

The family highlighted in the first case study was at a detention camp for two years immediately after the civil war ended in 2009 and then resettled in Jaffna, Sri Lanka, in a small, mud hut on a half-acre of land with fertile soil. While local and international voluntary relief organizations provided some supplies and materials to meet their immediate demands, initially, the five-member family had no assets and no support for any livelihood. As such, the family was looking for possible income generating activities and were struggling to find employment opportunities. The Sri Lankan army supported this family,

(a)

(b)

(c)

(d)

Photos 8.1a–8.1e (Continued)

(e)

Photos 8.1a–8.1e Selection of home gardens from northern Sri Lanka using low-cost inputs for crop cultivation

Source: Gunasingham Mikunthan and D. H. Galhena Dissanayake.

providing them with agricultural resources and manpower to establish a home garden in their half-acre land (Photos 8.1a to 8.1e).

The male head of the household was mentally and physically sick during the early days of their resettlement in Jaffna; therefore, the women of the household were involved in growing several crops within their small parcel of land. Soldiers volunteered their physical labor to establish the garden and introduced home gardening techniques commonly adopted in southern parts of Sri Lanka such as cultivation of flowering plants among food crops. They used locally available materials, including recycled items. For example, plastic and iron shell cases leftover from the war were made into containers to grow different kinds of leafy vegetables (see Photos 8.1a–8.1e). Through funds from the National Thematic Research Program (NTRP) of the National Science Foundation the family obtained necessary materials and training from the Faculty of Agriculture at University of Jaffna and the Department of Agriculture to strengthen their home garden. Periodically, assessments were made by the staff members engaged in the research program, and necessary support was given to the family to expand their income generating activities. This family slowly established their home garden with a diverse range of crops, including vegetables (e.g., tomato, chili, brinjal, okra, cabbage, beans), fruit crops (such as papaw, banana, mango, guava and pomegranate), leafy vegetables (*Centrella* sp., small *Amaranthus, Moringa, Sesbania* sp. *Pothina* and *Alternanthera* sp.), tubers/yams (e.g., elephant yam) and many medicinal plants, amounting to 82 species. Most of the products obtained from the home garden were primarily used for household consumption; any excess was sold later. The children found diverse food materials, especially fruits obtained from their home gardens.

The Sri Lankan government through its military units in Jaffna Peninsula and the Department of Agriculture conducted contests for the best home garden. These contests were designed to motivate resettlers to engage in home gardening and the winning households served to demonstrate best practices. Such contests encouraged innovation and boosted the uptake of a wider range of home gardening activities, including the production of livestock and beekeeping. Some households learned how to produce vermicompost which they could then sell for additional income. Home gardening proved to be not only productive and financially rewarding, but it also helped to build community ties, an aspect pertinent for families to build social capital in resettled areas. The garden of the household highlighted in this case won the second prize in the contest in 2014.

Over time the family obtained a cow, two goats and 15 chicken, which enhanced their household income further. The family also kept two beehives and a vermicomposting unit within the home garden itself. The father of the family developed an interest and enthusiasm toward found growing crops and was keen to learn more about the different crops and methods of cultivation. Faculty of Agriculture of University of Jaffna provided him with the requested training to support the continuation of home gardening activities. The family also received a water pump from a non-governmental organization which helped them to extend cultivation and expand production onto leased land near their household. The family is now established and has received a house from a housing scheme project. As their income slowly increased, they also purchased a bicycle that the mother used to take her kids to school. The children are happy as their living standard improved substantially after their resettlement. Other families that participated in this home gardening initiative also have interesting experiences to share. Overtime, these families were able to escape hunger and poverty and reach a state of sustainable food security and livelihoods through the expansion of their home gardening activities.

Case study 2: double-trouble for tsunami victims of Northern Sri Lanka

A second family showcased here is living in Kudarappu village of northern Sri Lanka. Kudarappu is a fishing village on the northeast coast that was affected both by the 2004 tsunami and then by the civil war. The village is isolated, approximately 15 km from the nearby market, and therefore the villagers mainly live on the fish they catch, with hardly any vegetables or leafy vegetables in their

Photo 8.2 Typical house with no vegetation in the northeast coastal area of Jaffna Peninsula, Sri Lanka

Source: Gunasingham Mikunthan.

diets. Malnutrition is high in the northern province of Sri Lanka with a rate above 40%. Unfortunately, in Kudarappu the rate is even higher because of lack of diversity in the food they consume. Obtaining vegetables and fruits from the nearby market is very difficult due to the village's remote location, and cultivation of vegetable and fruit crops is nearly impossible on sandy soils. Therefore, a major task of recovery from both the tsunami and the war was to find strategies to make vegetables and fruits available within the village. The best interventions were to establish home gardens to help diversify diets. In this case, generating income to meet villagers' other requirements was a secondary priority.

As noted earlier, the cultivation of crops on the village's soil, which was sandy and of poor quality, was a difficult task (see Photo 8.2). Transforming the

(a)

(b)

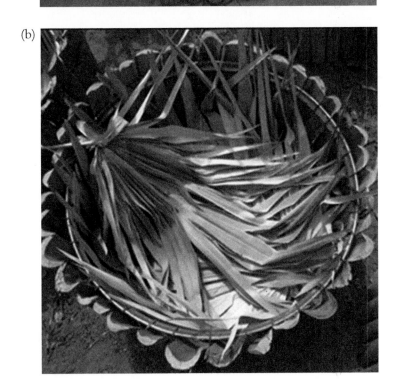

(c)

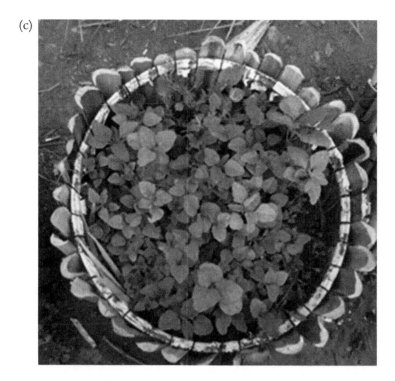

Photos 8.3a–8.3c Productive unit made with low-cost locally available materials and growing *Amaranthus*

Source: Gunasingham Mikunthan.

household within the available land into a greener environment and establishing a garden that could provide necessary food were the challenges. The Faculty of Agriculture at the University of Jaffna based on the agroecology of the area and developed a simple technological solution to facilitate crop production in sandy soils. The research team designed a productive cage with locally available waste materials and used it to establish plants in sandy soil. It was made from dried Palmyra petioles, two used bicycle tires and tire ropes (peeled from the used tires of heavy vehicles; see Photos 8.3a to 8.3c). Reusing these old bicycle tires and rims that can be breeding grounds for mosquitoes also helped to reduce the spread of dangerous diseases such as dengue, Japanese encephalitis and chikungunya. The cage was created to confine organic waste. Without this protective cage, organic matter would be been blown away by strong coastal winds. A second purpose of the cage was to prevent birds, reared within the household, from damaging the growing seeds. The innovative and inexpensive unit then allowed for the growing of vegetables and leafy vegetables around household and made them easy to identify. The research team also found that sprinkling 10 ml of distillery spent wash (a microorganism-rich, liquid waste

Photo 8.4 Training program demonstrating how to prepare a trench to preserve moisture
Source: Gunasingham Mikunthan.

from the distillery unit brewing Palmyra arrack from the toddy obtained from the inflorescence) into the cage facilitated the decomposition of the household plant and animal wastes. Thus the cage was an effective, easily implemented solution to overcome the pervasive problem of sandy soils inhibiting the cultivation of horticultural products.

Households in this area were supported by an international non-governmental organization (INGO) called ZOA from Australia as well as the research team from the University of Jaffna which helped to train villagers in home gardening around their households. Because most men were employed in fishing, the training programs targeted women. The university team taught village women how to produce the cages (Photo 8.4) and found that they quickly became

(a)

(b)

(c)

(d)

Photos 8.5a–8.5d Home gardening produces a greener environment with diverse plants

Source: Gunasingham Mikunthan and D. H. Galhena Dissanayake.

highly skilled at replicating them. ZOA supplied essential seed materials and provided education on crop growing techniques as well as the use of waste as organic manure for the area's sandy soil.

The characteristics of the household portrayed in this case is represents other homes in coastal fishing communities in northeastern Sri Lanka. This family had six members but lost three of them to the tsunami. They subsequently lost their dwelling and household belongings and had to leave the area due to the mass displacement caused by the intensified war during 2009. After the war, they returned to Kudarappu village where they had to reestablish their lives and livelihoods. Their main occupation had been fishing, but unfortunately, the family lost male members who were the primary income earners. The remaining male head of the household became sick and could not do any hard work. Thus, after resettling, they were only able to construct a small, Cajun house and had to survive on relief materials given to them by INGOs and the Government of Sri Lanka. Due to their poverty and lack of nutritious food in the areas, family members suffered from malnourishment.

As the male family members were either deceased or infirm, the woman of the household sought assistance in establishing a home garden to generate food and income. Through her participation in a survey of resettled families, she was able to request for support. She received training to use the techniques described earlier. The intervention was successful: today many plants are established in the household and greenery is seen in the sandy soil. Perennials such as *Moringa* and *Aloe vera* are growing well in this soil since they have symbiotic association of mycorrhiza in their roots. The garden also has onion and gourds as well as chili, tomato, okra, mango, banana, cashew and pomegranate. In particular, healthy, leafy vegetables such as *Moringa, Amaranthus, Pothina, Alternathera, Sesbania* and *Centella asiatica* are now grown and consumed by the household.

As a result of the home gardening intervention, the family has established a small shop within the house and started selling excess produce to the neighbors. With the environment full of diversified plants, their diet is now enriched with nutritious food products. Given the continued interest shown by the household to improve and expand their home gardening activities, when the village received electricity the research program provided the household with an electric water pump, a tank, accessories and other inputs to motivate them further. The woman's garden has also had a demonstration effect: neighbors visit her house and learn how to grow plants in sandy soil (see Photos 8.5a to 8.5d). Thus the model developed by the University of Jaffna serves as a model for home gardening in other coastal areas of the country.

Conclusions

Violent conflicts and natural disasters, such as the 30-year civil war and the 2004 tsunami in Sri Lanka, can have dire consequences for food security and the livelihoods of affected populations, particularly those in marginalized communities. As the case studies from northern Sri Lanka show, crises directly impact food availability and access and as a result, impair food utilization. Both family case studies illustrate the need to help crisis-affected households overcome malnourishment and poverty by making a diverse range of nutritious foods accessible while also enabling them to earn for comfort living. As Korf and Bauer (2002) note, interventions to overcome the adverse consequences of crises must incorporate five aspects of recovery – physical, environmental, economic, social and political. As these Sri Lankan case studies demonstrate, interventions by governmental organizations and NGOs tackled the multiple effects of the tsunami and the war. Physical aspects of rehabilitation included the reconstruction of infrastructure (e.g., roads, hospitals, schools), resettling of displaced families and providing them with land and sometimes housing. In addition, meeting these physical requirements, interventions also began to address environmental, economic needs, social and political facets of recovery. Home gardens proved to be an effective intervention to meet many of these needs.

Home gardening promoted environmentally friendly cultivation practices: for instance, the home gardening intervention in villages of northern Sri Lanka enabled families to recycle organic wastes for manure and use by-products of the war (such as ammunition casings) to construct garden cages and containers and improve productivity. The intervention also promoted the cultivation of toxin-free foods to enhance families' health and nutritional well-being. Economically, home gardens initially were intended to provide households with a measure of food self-sufficiency; however, some home gardeners were able to barter or sell excess produce for additional income, thus enhancing their livelihood capacities and household food security. This dimension of the program was particularly useful for female-headed households and families with disabled family members who could not secure employment outside the home. Socially, home gardens helped to build ties across the community, as families exchanged both produce as well as knowledge about best practices. They also provided an important source of nutritious foods that were generally unavailable in local markets, as in the case of the fisheries household whose diet was traditionally limited. Finally, the home gardening initiative undertaken by the Government of Sri Lanka with support rendered by the military wing at Jaffna, the Department of Agriculture and NGOs and INGOs worked to build trust, which had been undermined during the many years of war, between various actors concerned with redevelopment efforts. The work begun under this program continued in the years following the end of the war through the expansion of extension services and available public services. Through the support given by public and private organizations, crisis-affected families participating in the home gardening programs in Sri Lanka were able to access inputs and training that enabled them to withstand production constraints, generate income and sustain home gardening activities. Thus, home gardening proved to be an effective intervention to build resilience in crisis-affected families and communities, allowing them to adapt to changes and constraints caused by crises. The innovation developed to establish home gardens in sandy soils can be replicated or adapted in other coastal areas to cultivate vegetables and fruits using recycled materials. The experiences from northern Sri Lanka showcase the versatility and adaptability of home gardening in diverse situations and its capacity to diversifying diets and income sources. The United Nations (UN) declared the year 2014 as the 'International Year of Family Farming'. Gardening at home by default is family farming in which household members partially or totally supply labor and cultivate in a small landholding. Despite being an adaptable and versatile low-cost solution to address hunger and poverty, home gardening has received very little recognition and support. In fact, most development initiatives push for land consolidation and large-scale farming and overlook the vital contributions of home gardens for food and nutritional security and livelihoods as well as biodiversity.

The recent advent of the Corona virus (COVID-19) pandemic in 2020 has exacerbated and created new challenges on food systems. There are widespread concerns amongst development groups on how to sustain domestic food production; maintain safe food value chains; and keep markets adequately supplied amidst the crisis. Not only that, on the demand side with millions of people losing jobs and unemployment rates escalating rapidly, nations have to device strategies to make food socially, physically, and economically accessible yet maintain a diverse supply. As the economic consequences and restrictions to population movement due to the pandemic unfold, households are losing their ability to purchase food and other essential consumer products. If the situation continues to deteriorate it will be a while before food systems become fully operational and economic activities return to normal.

The prevailing circumstances are triggering renewed interest in home gardening as households are reverting to practices that could strengthen their resilience to exogenous threats. Home gardening is a flexible form of small-scale agriculture that can fit and be adapted to various socioeconomic contexts. As Chapter 10 of this book demonstrates, even the landless can practice home gardening. Families can engage in home gardening and grow diverse crops at their homestead to feed the family and barter or sell any surplus for additional income with minimal inputs and investment of their energy. As such, home gardens can be the building blocks of nations' food and nutritional security and directly contribute to the sustainable development goals (SGDs) 2 (Zero hunger), 3 (Good health and well-being) and 1 (No poverty) while complementing other SDGs.

With the launch of the 'UN Decade of Family Farming' in 2019, optimism is raised that due attention to this neglected area of the local food system and put in place policies and programs to support home gardening. As the needs of home gardeners across the globe are similar innovative approaches, and best practices can be shared and adapted to diverse cultural, economic, and socio-political contexts.

Note

1 Where there is "availability at all times of adequate world food supplies of basic foodstuffs to sustain a steady expansion of food consumption and to offset fluctuations in production and prices" (United Nations, 1975).

References

Bandarin, F., Hosagrahar, J. and Albernaz, F.S., 2011. Why development needs culture. *Journal of Cultural Heritage Management and Sustainable Development*, 1(1), pp. 15–25.

Boly, A., 2013. *Building resilience to address the root cause of food security*. [Online] Available at: www.ifrc.org/en/news-and-media/meetings-and-events/other-events/global-platform-on-disaster-risk-reduction-2013/building-resilience-to-address-the-root-cause-of-food-security/ [Accessed 12 July 2019].

Brinkman, H.J., et al., 2010. High food prices and the global financial crisis have reduced access to nutritious food and worsened nutritional status and health. *The Journal of nutrition*, 140(1), pp. 153S–161S.

Cannon, T., 1994. Vulnerability analysis and the explanation of "natural" disasters. *Disasters, Development and Environment*, 1, pp. 13–30.

Chambers, R. and Conway, G.R., 1991. *Sustainable rural livelihoods: Practical concepts for the 21st century*. Discussion Paper 296. Brighton, UK: Institute of Development Studies.

Collier, P., 2009. Post-conflict recovery: How should strategies be distinctive? *Journal of African Economies*, 18(suppl_1), pp. i99–i131.

Comfort, L.K., Boin, A. and Demchak, C.C. eds., 2010. *Designing resilience: Preparing for extreme events*. Pittsburgh, PA, USA: University of Pittsburgh Press.

Department for International Development, 2011. *Defining disaster resilience: A DFID approach paper*. London, UK: Department for International Development.

Devereux, S. and Edwards, J., 2004. Climate change and food security. *Bulletin: Institute of Development Studies*, 35, pp. 22–30.

Food and Agriculture Organization of the United Nations, 2001. *The state of food insecurity in the world 2001*. Rome, Italy: Food and Agriculture Organization of the United Nations.

Food and Agriculture Organization of the United Nations, 2009. *The state of food insecurity in the world 2009*. Rome, Italy: Food and Agriculture Organization of the United Nations.

Food and Agriculture Organization of the United Nations, the International Fund for Agricultural Development, the United Nations Children's Fund, the World Food Programme and or the World Health Organization, 2019. *The state of food security and nutrition in the world 2019*. Rome, Italy: Food and Agriculture Organization of the United Nations.

Galhena, D.H., Freed, R. and Maredia, K.M., 2013. Home gardens: A promising approach to enhance household food security and wellbeing. *Agriculture & Food Security*, 2(1), p. 8.

Holling, C., 1973. Resilience and stability of ecological systems. *Annual Review of Ecology and Systematics*, 4(1), pp. 1–23.

International Fund for Agricultural Development, 2010. *Rural poverty report 2011: New realities, new challenges: New opportunities for tomorrow's generation*. Rome, Italy: International Fund for Agricultural Development.

International Institute for Strategic Studies, 2019. *Armed conflict database*. [Online] Available at: www.iiss.org/publications/armed-conflict-database [Accessed 12 October 2019].

Korf, B. and Bauer, E., 2002. *Food security in the context of crisis and conflict: Beyond continuum thinking*. London, UK: International Institute for Environment and Development.

Landon-Lane, C., 2011. *Livelihoods grow in gardens: Diversifying rural income through home garden*. Rome, Italy: Food and Agriculture Organization of the United Nations.

La Vía Campesina, 1996. *The right to produce and access to land: Position of La Vía Campesina on Food Sovereignty*. Rome, Italy: World Food Summit.

Marsh, R., 1998. Building on traditional gardening to improve household food security. *Food, Nutrition and Agriculture*, 22, pp. 4–14.

Mattsson, E., Ostwald, M. and Nissanka, S., 2018. What is good about Sri Lankan homegardens with regards to food security? A synthesis of the current scientific knowledge of a multifunctional land: Use system. *Agroforestry Systems*, 92(6), pp. 1469–1484.

McConnell, D.J. and Dharmapala, K.A.E., 1973. The economic structure of Kandyan forest garden farms: Small forest garden farms in the Kandy district of Sri Lanka. *Farm Management Diversification Report*, (7).

Mendez, V.E., Lok, R. and Somarriba, E., 2001. Interdisciplinary analysis of homegardens in Nicaragua: Micro-zonation, plant use and socioeconomic importance. *Agroforestry Systems*, 51, pp. 85–96.

Mukherjee, A., Upendranadh, C., Jones, B. and Roncarti, M., 2010. *Social access and social protection for food security in Asia Pacific*. Bogor, Indonesia: Centre for Alleviation of Poverty through Sustainable Agriculture-Economic and Social Commission for Asia and the Pacific.

Niehof, A., 2004. The significance of diversification for rural livelihood systems. *Food Policy*, 29(4), pp. 321–338.

Niehof, A. and Price, L., 2001. *Rural livelihood systems: A conceptual framework* (1st ed.). Wageningen, The Netherlands: WU-UPWARD.

Niñez, V.K., 1984. *Household gardens: Theoretical considerations on an old survival strategy*. Lima, Peru: International Potato Center.

Odebode, O.S., 2006. Assessment of home gardening as a potential source of household income in akinyele local government area of Oyo State. *Nigerian Journal of Horticulture Science*, 2, pp. 47–55.

Olinto, P., Beegle, K., Sobrado, C. & Uematsu, H., 2013. *The state of the poor: Where are the poor, where is extreme poverty harder to end and what is the current profile of the world's poor?* Washington, DC, USA: The World Bank.

Pehu, E., Lambrou, Y. and Hartl, M., 2009. *Gender in agriculture sourcebook*. Washington, DC, USA: The World Bank, FAO, IFAD.

Racioppi, L. and Rajagopalan, S. eds., 2016. *Women and disasters in South Asia: Survival, security and development*. Abingdon, UK: Routledge.

Smith, M., Pointing, J. and Maxwell, S., 1993. *Household food security: Concepts and definitions: An annotated bibliography*. Brighton, UK: Institute of Development Studies, University of Sussex.

Teodosijević, S., 2003. *Armed conflicts and food security*. Rome, Italy: Food and Agriculture Organization, ESA Working Paper No. 03–11.

United Nations, 1975. *Report of the world food conference, Rome 5–16 November 1974*. New York, USA: United Nations.

United States Agency for International Development (USAID), 2014. *Building climate change resilience and food security program.* [Online] Available at: https://2012-2017.usaid.gov/kenya/fact-sheets/building-climate-change-resilience-and-food-security-program [Accessed 1 August 2019].

World Food Summit Plan of Action, 2002. *Food, security, justice and peace.* Rome, Italy: Food and Agriculture Organization of the United Nations.

9 Complementarity between the home gardening and livestock production systems in Nepal

Ramjee Ghimire and Nanda Joshi

Background

Nepal is a small landlocked Asian country spanned across 56,826 square miles area ranging from 60 masl (meters above sea level) to the tallest peak in the world at 8,848 masl. Crops are grown from the 60 masl to 4,700 msal (Joshi et al., 2018). Agriculture provides livelihoods for 68% of Nepal's population, accounting for 34% of the GDP (United States Agency for International Development (USAID) 2019). Yet Nepal has been facing food insecurity, malnutrition and hunger for many years now. Thirty-six percent of Nepali children under the age of five years suffer from chronic malnutrition or stunting, which causes debilitating effects such as blindness, brain damage and infectious diseases, which can result in lifetime damage (USAID, 2019). Also, maternal health and deaths remain serious concerns among women. With per capita gross national income of $960, human capital index of 0.49 and human development index of 0.574 (World Bank, 2019), Nepal has a long way to go to attain progress and prosperity. Despite these shortfalls, Nepal has a unique social, ecological and geographical set-up that can be harnessed to foster economic growth.

Major agricultural systems in Nepal

Nepal follows a mixed farming system and is rich in agrobiodiversity. FAO (2020) explains Nepal's farming system and agrobiodiversity as dominated by farmers that practice mixed farming systems integrating livestock, horticultural crops and cereals. Nepal is richly endowed with agrobiodiversity. Rice, maize, millet, wheat, barley and buckwheat are the major staple food crops. Similarly, oilseeds, potato, tobacco, sugarcane, jute and cotton are the important cash crops whereas lentil, gram, pigeon pea, black gram, horse gram and soybean are the important pulse crops. Farmers in Nepal grow several fruits and vegetable crops. Some important ones are apple, peach, pear, plum, walnut, orange, lime, lemon, mango, litchi, banana, pineapple, papaya, cucumber, lady's finger (okra), brinjal (eggplant), pumpkin and several leafy vegetables. Nepal is also famous for orthodox tea, large cardamom, turmeric and ginger. Most Nepalese farmers diversify their crops production to hedge against erratic and uncertain weather and other unfavorable agronomic conditions. Many of these crops are grown on lands where homes are; such systems are called *Ghar Bagaincha* or home gardens. Home gardens operate year-round and serve many purposes, the major one being the supply of food. Nepalese families, especially in rural areas, cannot imagine life without home gardens, thus they value home gardens highly.

Nepal's agriculture is dominantly of the mixed type. Most families raise livestock at home and they have some farmlands for the cultivation of cereals, vegetables, commercial and other crops. Livestock is one of the important sources of cash income of the farm households in Nepal. Due to the diverse topography and climates, many diverse livestock species are raised in Nepal (Table 9.1). Half of the fowl are backyard poultry in Nepal (Department of Livestock Services, 2018; Kattel, 2016), which are mostly of indigenous breed. Indigenous poultry has many merits including but not limited to the fact that they are resistant to many contagious diseases, they fetch higher prices and can be raised in small area with minimal inputs. Pigeons and ducks are other popular poultry mostly in rural and peri urban areas. Importantly, Nepal has one of the highest number of buffaloes per unit area of the land in Asia (Abington, 1992).

Livestock and poultry are grown for meat, milk, eggs, wool, manure and draft purpose. Most livestock producers in Nepal, particularly in hills and mountains (rural areas), are smallholder subsistence farmers. Most of their produce are consumed at home. Livestock products such as meats, milk and eggs are a rich source of dietary proteins. Khanh et al. (2006) states that smaller the farm size and farm economy, the higher is the contribution of livestock on livelihoods; this scenario applies to Nepal. The cash needs of the farm families are mainly met through the sale of milk, yoghurt, cheese, ghee, *chhurpi* (traditional cheese), meat, eggs, live animals and poultry. Generally, farm families in mountains raise yaks or *chauris* (Himalayan breed of cow) and sheep. In the hills, cows, sheep, goats and poultry are raised. Buffalos, cows, goats and poultry are reared in the Terai. Poultry husbandry is an emerging enterprise in the Terai and hills. More than 70% of farming households raise livestock and poultry in Nepal. These livestock are raised at and in sheds close to farmers' homes. Many of the products or by-products of livestock and poultry farms are used in home

Table 9.1 Livestock production in Nepal

Species	Number	Household owning these livestock
Cattle	6,430,397	2,280,542
Buffalo	3,174,389	1,668,820
Yak/*chauri*	48,865	6,235
Sheep	612,884	98,464
Goat	11,225,130	2,463,253
Pig	870,197	477,984
Fowl	70,007,151	–
Duck	394,775	–

Source: Department of Livestock Services, 2018; Kattel, 2016; Ministry of Livestock Development, 2017.

gardens in terms of many home garden products, and by-products are utilized for livestock production, therefore in Nepal, home gardens and livestock always go together. However, limited access to improved seeds, new technologies and market opportunities is hindering agricultural production and productivity in Nepal (USAID, 2019). The USAID reports that declining agricultural production has depressed rural farming economies and increased widespread hunger and urban migration.

With this background, this chapter discusses home gardens and livestock production and their interrelationships in Nepal. Let us begin with the definition of home gardens. Home gardens are defined as intentional associations of trees, shrubs, herbaceous crops and/or animals within the homestead boundaries of rural and urban families, which are rich in agrobiodiversity (Fernandez and Nair, 1986). Home gardens provide a wide range of products, including but not limited to food, ornamentals, medicines, firewood, timber, construction materials, fodder and others, for home use, gifts, exchange and trade (Nair, 1991).

Importance and attributes of home garden in Nepal

In Nepal, a home garden refers to the traditional land use system around a homestead, where multi-purpose trees, shrubs, herbs, annual and perennial agriculture crops, spices, medicinal, ornamental plants and livestock are managed by family members to fulfill their multiple requirements (Shrestha et al., 2002). About 72% of total households of Nepal have been maintaining home gardens occupying an area of 2%–11% of total land holdings (Gautam et al., 2004).

Pulami and Paudel (2004) have discussed many importance of home gardens in Nepal. Home gardens are an easily accessible and rich source of green vegetables that provide dietary nutrients. Specifically, women, the elderly, children and people of low-income families have difficulty finding quality foods and cannot afford to buy foods from market centers. In such cases, home gardens serve as a key source of food for major dietary requirements (Photo 9.1). Home gardens are source for cash income. There are families who sell fruits, vegetables, greens or other home garden products to their neighbors or to nearby shops and earn cash income that they use to acquire other necessities. Home gardens help make families self-dependent as they grow their own vegetables and fruits and raise their livestock. There are other important benefits from home gardening to households in Nepal.

Home gardens in Nepal are one of the important centers of experimentation, species domestication and crop improvements. Being rich source of diverse crops, for example, over 254 species of 197 genera belonging to 76 families of crops in home gardens in eastern and central hill and Terai regions, home gardens in Nepal represent an important reservoir of diversity of plant species and have immensely contributed to the maintenance, promotion and in situ conservation of plant genetic resources (Subedi et al., 2006).

Photo 9.1 Various vegetables grown in a home garden

Source: Ramjee Ghimire.

Import substitution and unabated supply

Nepal imports a large chunk of vegetables and fruits every year from India and this trend is ever increasing. The Department of Customs (DoC) record shows that vegetables worth INR 14.77 billion were imported in the first half of fiscal year 2018–2019 (as cited in *Himalayan Times*, November 25, 2019).

Home gardens could be extremely beneficial to minimize such dependencies on imported foods. Connected to this is the unabated supply of fruits, vegetables and other edibles from the home garden. There are days in summer when transportation does not operate due to landslides, floods and so forth. Anecdotal evidence indicates food deficits in the hills and mountains is attributed to difficulty in transportation. Therefore, local production in the home garden would serve as the remedy to such blockades. Food products from home gardens help avert possible shortages in food supplies due to such natural adversities and help minimize the risk of food insecurity for rural populations.

Source of cereals, fresh produce and fodder

As can be seen in atypical homestead in western Nepal home gardens are popular for growing cereals such rice, millet and maize in Nepal. Once harvested cereals are stored in locally made grain baskets. Owners mill those cereals, process them and consume as needed.

Rice is the staple food in Nepal which is usually served with lentil soup, curry, (green) vegetables and pickle. Some households sell their surplus cereals for cash. Home gardens also provide fresh and healthy food products. Some crops grown in home gardens are special crops which are either more aromatic, or fetch more values, or are used to prepared special feasts in festivals. Home gardeners choose to grow such crops because they can monitor them regularly and provide intensive cares. Home garden products are dominantly organic, free of any insecticides and pesticides. There is a common practice of growing fodder trees on home gardens. Such fodder trees serve as the source for green feeds for livestock (Photo 9.2).

Employment generation

Several thousand youth go to Gulf countries, Malaysia and other countries in search of jobs every year (Government of Nepal, 2018). If properly trained, they can pursue home gardening activities for income generation. Some home garden products such as turmeric, ginger, coffee are processed, mostly by women and such products fetch high values in the local markets.

Social empowerment

Home gardens not only supply foods but are also status symbols of power and prosperity. Field research reveals that households of so-called lower castes do not face the same discrimination that others of the same caste face, if they have well-maintained home gardens. Linked to this is the fact that members of households with home gardens are better off with regard to nutrition and health as they have better access to nutrient rich foods such as fruits, vegetables, greens and livestock products, and they can share excess products (e.g., banana) with their neighbors (Photo 9.3).

Gender (women) empowerment

Nepal has dominantly patriarchal culture. Women do not enjoy the same access to and control over resources that men do. However, women in the households with home gardens are more empowered socially, economically and

Photo 9.2 Fodder production in a home garden

Source: Ramjee Ghimire.

nutritionally (Sthapit et al., 2006). Home gardens provide women and children have easy access to preferred foods, such as cereals, fruits and vegetables (Sthapit et al., 2006). As reported in Talukder et al. (2006), home gardening interventions implemented in Nepal by Hellen Keller International corroborate that the proportion of pregnant and lactating mothers and the children under five that consumed fruits and vegetables increased considerably as a result of home gardens. Furthermore, in the communities where interventions were made, consumption of egg was doubled in both the mothers and children under five. Talukder et al. (2006) further elaborates that:

> the percentage of households earned money by selling poultry and eggs in two months period increased from 18% to 58% and median amount of money earned was from NRs 188 to NRs. 322. HKI homestead food

Photo 9.3 Fruits grown in a home garden in western Nepal
Source: Ramjee Ghimire.

production program increased both production and consumption of micronutrient rich foods including plant and animal sources. It further helped to increase the quality of the household's diet. Homestead food production also increased family income that increases household food and nutrition security.

(p. 27)

Recreation and aesthetic beauty

Home gardens are avenues for recreation as family members spend their time working in them, tending to crops and animals, appreciating the products (fruits, flowers, etc.) and importantly enjoying natural phenomena, for example, biology, diversity and characteristics of plants and animals. Households grow many different flowers in their home gardens that they use to beautify their homes and as well for cultural purposes. Nepal imports over NPR 15 million worth of roses on Valentine's Day (*Indian Express*, February 9, 2019) and NPR 110 million worth of flowers during the Tihar festival alone (Nepali Sansar; October 31, 2018). Flowers grown in home gardens can significantly help minimize such imports while providing economic incentives to households.

Family cohesion

In home gardens, family members irrespective of their ages work together and contribute to the extent possible to make home gardens a success. Of course, their roles differ by their age, expertise, experience and power. Home gardens are a place for deliberation and reflection on day-to-day affairs as well as a venue for critical review of agronomic, animal husbandry and other production practices.

Learning platforms/laboratories and technology adoption

First steps to farming are made through home gardens. Especially the younger family members watch and learn from their parents and other elderly members of the household. Children and youngsters are often asked to help or participate alongside older family members to perform home gardening activities. This engagement helps them acquire skills and knowledge on how to work the soils, sow seeds and cultivate crops, care for crops and animals, recycle household organic waste and farmyard manure and when to harvest. Home gardens also serve as places for experimentation. The household may test new varieties, do simple breeding and selection, and try out new methods of propagation, cultivation and post-harvest techniques. These interactions help build relationships and family ties. Further, they serve as models for knowledge sharing and demonstrations. As such, the best practices of one home gardener can be shared with others using the home garden as a demonstration site (Photos 9.4a and 9.4b). Home gardeners often take pride in disseminating the results of their gardening experiments, the proven technologies and/or the outputs from their trials with each other.

Utilization of household and garden wastes

Reclining of household and kitchens waste, leftover husks after thrashing rice, wheat, maize and so forth are disposed of and decomposed. Composting of these waste products can provide households with fertilizer rich in organic matter with little to no costs. Many rural households in Nepal use firewood for cooking. Some of the firewood come from trees and crop residues (maize stover and cobs) from the home gardens.

(a)

(b)

Photos 9.4a and 9.4b Home garden with tomato and flowers
Source: Ramjee Ghimire.

Biodiversity conservation

Home gardens serve as the source for green vegetables, fruits, herbs, cereals and flowers, among others. Most home gardens are multi-level cropping systems with three to four tiers of perennial, annual and seasonal plants. They include annual and peri-annual small trees and shrubs (herbs, greens, vegetables, tubers, coffee, etc.), vines (yams, fruits, etc.), peri-annual trees (mangoes, jack fruit, oranges, guava, banana, etc.). Vegetables grown in home gardens differ by season, altitude or locality. There are over 131 species of plants found in home gardens in Nepal (Sunwar, 2006). A study by Khanal et al. (2019) in Nepal recorded 106 species of 38 families that include 3 commercial crops, 7 cereals, 19 fruits, 21 medicinal plants, 5 spices, 11 fodder/trees and 42 vegetables.

How do home garden and livestock complement each other?

Livestock production is an integral part of home gardens in rural Nepal (Photo 9.5) (Sthapit et al., 2006). Livestock and poultry provide manure (farmyard manure) and urine, which are rich in nitrogen and other nutrients that plants require for growth. This has long been in practice in Nepal, especially in the high hills and mountains where herds are big. The livestock producer let their livestock herds/flocks roam and graze in their home gardens and in farmlands for couple of days or even weeks. The motivation behind such practice is to fertilize farmlands and in return allow livestock to graze there. This reciprocity has been going on for generations (Tulachan and Neupane, 1999). With the cost of fertilizer inputs on the rise, the use of urine as a nutrient supplement for crops is gradually increasing.

Home gardens also provide forage, fodders, hays and other leftover greens that livestock can eat. Home gardeners have been using leftover forage and fodder and beddings from livestock sheds (Photo 9.5) to prepare compost and as mulch. Leftover stovers, leaves and roughage from home gardens are used as bedding for livestock. By-products from fodder production such as stems (broom grass, etc.) are utilized for staking tomato and other similar plants by home gardeners. Furthermore, livestock (oxen) are used to plow farmlands, including home gardens.

Livestock systems in Nepal are rich in genetic biodiversity. Pigs are an integral part of mixed farming system, specifically among ethnic tribes (Gautam et al., 2006). In poultry alone, there are many indigenous breeds, for example, Shakini, Ghanti khuile and Pwankh Ulte are such popular breeds (Kattel, 2016).

Williamson and Lyons (2013) offer many benefits, co-benefits and synergies of integrating livestock and home gardens. In such a system, many naturally introduced and naturally occurring plants, animals and microorganisms compete with, assist each other and co-exist. Growers of such farms attempt to maintain and take advantage of naturally occurring relationships instead of

Photo 9.5 Livestock in home gardens in rural Nepal
Source: Ramjee Ghimire.

fighting or destroying them and their efforts are to combine livestock and crops in a way that is mutually beneficial. They further add that "Layering animals and vegetables can create positively reinforcing systems while producing more food through increased fertility and layered production systems" (p. 1).

Issues and challenges when integrating

Disease vulnerability

Nepalese smallholder farmers pursuing integrated or mixed farming keep livestock and poultry very close to their residences. Some buildings or houses are even shared by humans, livestock and poultry. Furthermore, livestock and poultry sheds are often not clean. Therefore, it is always challenging to keep human being safe from animal-borne diseases such as zoonotic diseases.

Underutilized genetic diversity

Nepal is yet to complete genetic profiling of its native livestock and plant species. Genetic improvement through selection combined with improved husbandry practices greatly help increase livestock productivity. There are poultry breeds (Photos 9.6a and 9.6b), which are disease resistant, perform well in semi-intensive management and fetch good prices, yet these resources are not used optimally to increase livestock production in home gardens.

(a)

(b)

Photos 9.6a and 9.6b Local fowls in Nepal are an integral part of home gardens
Source: Ramjee Ghimire.

Rapid urbanization

With the expansion in road networks, electricity and telecommunication, there is increase in urbanization in Nepal. Processed foods and goods supplied in attractive packages are more accessible now than before. Among present-day consumers there is growing attraction for such foods and goods than local or indigenous foods (grown in home gardens) and goods. There is also a lack of awareness among urban dwellers about the nutritive value of local foods. Over-reliance on packaged and processed foods can erode the nutritional status of the population and result in adverse environment consequences.

Irrigation in winter

Winter water shortages are a big issue for Nepal; therefore, it is difficult to maintain home gardens during winter, or the dry season. However, realizing this constraint, a few households have started storing rainwater in cement tanks or plastic pits and use that water to irrigate crops in the gardens during periods of water scarcity. As a risk averting practice, in many households the place to wash utensils is located next to or above the home garden plots so the wastewater (even at minimal quantities) drains to the home gardens helping to irrigate plants.

Climatic changes and effects on environment

Most home gardeners and livestock producers are smallholders with little or no formal education, although agricultural assets, including the home garden and livestock, are being inherited and passed on to the next generation. Many of them are unaware of the ongoing climatic changes; those that have some awareness are not motivated or are unable to adapt to those changes.

Poor planning and layout

A majority of the livestock sheds or poultry houses and home gardens in rural Nepal are poorly designed (Photo 9.7). Livestock sheds lack ventilation and there are no proper waste disposal systems in place. Poor disposal of dung and livestock urine is hazardous to human and animal health and well-being and increases vulnerability and spread of diseases.

Encroachment of livestock during winter and dry months

In Nepal, 80% of the rain is received during monsoon or summer (July–September) period, but during winter it is very dry (Photo 9.8). This rainfall pattern has a huge impact on home garden and livestock production systems. Since most farmlands remain fallow (Photo 9.9) and livestock are let loose for grazing during winter, oftentimes those livestock encroach on home gardens and feed on the plants reserved for human use. Homeowners find it difficult to protect their home gardens from such encroachment during winter. Those who manage to build fences around their home gardens are able to sustain their home gardens, but those that cannot face losses.

Lack of research, extension and training on better livestock production practices for home gardens

The government expends minimal effort to develop and promote home gardening and livestock production. The two programs are run in silos and there is not much integration between them. Consequently, assistance to home gardeners to develop their cropping, livestock and animal husbandry activities

Photo 9.7 Home garden and homestead in Nepal

Source: Ramjee Ghimire.

lack coordination. However, efforts by some notable non-profit sector organizations such as Local Initiatives for Biodiversity, Research and Development (LI–BIRD) have being instrumental in supporting home gardening projects.

Way forward

As noted earlier, there is not much research or training to improve and support livestock production in home gardens. So the key question is, how can these two sectors be developed to improve productivity and people's livelihoods? Some recommendations are as follows.

Photo 9.8 Rural village scene during winter
Source: Ramjee Ghimire.

Systems approach

Any effort to develop and integrate home gardens and livestock should be examined and pursued through a systems approach. It means identifying subsystems associated with home garden and livestock systems. It's also important to understand the interactions between these subsystems, for example, how the human and social sub-systems interact with the environmental and/or economic sub-systems. This understanding is key to any planning, research and development of interventions to stimulate home garden and livestock development.

Education

Growers, consumers and other players who are involved in value chain of livestock and home gardening should be educated about the importance of home

Photo 9.9 Fallow farmland in winter
Source: Ramjee Ghimire.

garden and livestock and their symbiotic relationships, contribution of these sectors to attaining household food and nutritional security, biology of flora and fauna associated with these systems, and their sustainable management. School children should be taught about the importance of integration of home gardens and livestock systems so that they can share their knowledge with their parents and other family members.

Extension

Extension efforts employing local mobilizers such as clubs (women, youth), agroforestry, farmers, self-help groups and cooperatives in languages understandable to local populations is paramount to advancing livestock production activities in home gardens and maintaining their continuity. Since a large majority of people in Nepal now possess smartphones, it will be worthwhile to use this widely used technology to disseminate information that will help home gardeners to manage their cropping and livestock processes better. For crops, the best time to sow vegetable seeds/seedlings, major pests and control and containment mechanisms, market prices of major home garden products and so forth. For livestock, disease outbreaks and epidemics in the region and control measures, information related to livestock feeding and management, breeding, housing and immunization against major diseases, zoonotic diseases and their prevention can be shared via smartphones.

Biodiversity

It is timely to develop and update the biodiversity and species profile of home gardens and livestock systems in Nepal. This will help identify economically and environmentally beneficial species and promote those species as well as conserve those genetic materials.

Gender

The women are nearly inseparable from livestock and home gardening systems in Nepal. However, there is little acknowledgement of their contributions to both home gardening and livestock raising. As such, they face constraints socially and economically hindering their ability to sustain integrated home gardening activities and increase productivity. Thus empowering women, improving their access and control over resources – land, technologies, inputs as well as enabling them to make decisions – production, consumptions and marketing of their products can greatly enhance their own well-being and of their family.

Value chain

Developing and/or strengthening home garden and livestock production–related value chains are essential. Efforts should be made to add value to such products through processing so that home gardeners can fetch a better price. This entails efforts to provide processing technologies and infrastructure that would make storage, preservation and processing of raw products into high value products possible (e.g., kimchi from greens, cheese from milk).

Promotion and marketing

Proper coordination and innovative strategies are needed to promote home grown products. Marketability of home garden products will encourage home gardening, and consumers too will have access to a variety of local products that may not be available otherwise. Approaches similar to farmer's markets can provide economic opportunities for home gardeners as well as access to products by consumers.

Input supply

Home gardening is not resource intensive. However, improved and timely access to inputs (seeds, seedling, suckers, breeding bulls for semen, chicken, feeds, veterinary medicines, etc.) can greatly enhance productivity and reap benefits to the household.

Research

There is an urgency for research on home gardening. Some immediate research needs include changes to biodiversity profile of home gardens, effects of climatic change on home gardening and livestock production, impact of migration (internal and out-migration) on home gardening and livestock raising and identifying models (species, size, breeds, production systems, etc.) for sustainable integration of livestock into home gardens, among others.

Management in winter or dry season

Building makeshift fences can help protect home gardens from damage by livestock and poultry during the winter and dry seasons. It is also important to find ways to harvest and sustain an adequate water supply to irrigate the home garden. For instance, during summer or the rainy season water can be collected in cement or plastic tanks and utilized efficiently during the dry winter months.

Conservation and preservation and value addition

There is possibility of introducing new technologies to preserve products (e.g., greens) when they are in surplus, for example in the summer or rainy season. Those preserved can be consumed in winter.

Consumption

There is possibility of developing additional recipes (foods), which are more nutritious, from the home garden and livestock products and/or combination of those two product groups than what are currently available. Showcasing such foods in local festivals or meetings will be helpful to educate growers as well as consumers about those products and their nutritive values. Such strategies are particularly useful in tackling issues related to undernourishment and micronutrient deficiencies in rural areas.

Climatic changes

Home gardeners should be apprised and educated about the effects of climate changes on home gardening and livestock production systems and be equipped with the measures to adapt to those changes.

Keeping surroundings clean

Livestock sheds should be kept clean, free from dung, foul smells, flies, mosquitoes and other pests. Home gardens should also be free from weeds and have good drainage. Livestock sheds and goat or poultry houses should have proper ventilation, and the litter, manure/dung and urine should be properly disposed or treated for other household purposes.

Promotion in urban and peri-urban areas

Urban and peri-urban dwellers are changing their practices and they are less inclined to maintain home gardens. Though it will be difficult to raise livestock in urban settings, peri-urban dwellers can still pursue some livestock, specifically goats and backyard poultry. Market would not be the problems for them.

Monitoring and evaluation

It is strongly recommended that the integrated crop and livestock systems in home gardens be evaluated periodically to determine their productivity/outputs, outcomes and impacts to the household economy, individual health and nutrition and the environment. Any planning for home garden development must be informed of these results to ensure premeditated approaches are taken for their enhancement.

Promoting production

Most home gardens and livestock managed in rural Nepal are organic and there is no or very minimal use of agrochemicals – fertilizers, pesticides, insecticides and so forth. This should be further encouraged, and home gardeners should be informed of technologies and management practices that would avert the use of agrochemicals to manage pests and diseases.

Planning and layout

Home garden owners should be educated about designing an aesthetically appealing layout and growing a combination of plants that would make gardens useful, easy to access and work, and aesthetically pleasing. A combination of perennial, annual or seasonal flowers around the home gardens would make gardens appealing and colorful. Similarly, livestock sheds should be designed to make the collection of urine and manure easy and minimize the foul ammonia odor around the shed and house.

Policy

After all, whatever people and institutions do in a country are influenced by government policies. Therefore, it is strongly recommended that the government introduce policies to facilitate the integration of home gardens and livestock production. These policies should help promote conservation of native gene pools, timely research on emerging issues, help link growers to markets and offer program to educate stakeholders and so forth. Furthermore, investigation of livestock diseases and diseases and pests of crops are public services. Livestock diseases and plant diseases and pests are likely to affect a large number of livestock/poultry and crops that require expert services which are beyond the control of individual farmers. Home gardeners and livestock raisers demand assurance of investigation and timely control and containment measures of these cases.

References

Abington, J.B. ed., 1992. *Sustainable livestock production in the mountain agro: Ecosystem of Nepal* (No. 105). Food & Agriculture Organization. Available at: www.fao.org/3/T0706E/T0706 E00.htm#TOC

Department of Livestock Services, 2018. *Livestock statistics of Nepal 2016/17.* Available at: http://dls.gov.np/reportsdetail/1/2018/37423677/

FAO, 2020. *FAO in Nepal: Nepal at at glance.* Available at: http://www.fao.org/nepal/fao-in-nepal/nepal-at-a-glance/en/

Fernandes, E.C. and Nair, P.R., 1986. An evaluation of the structure and function of tropical home gardens. *Agricultural Systems,* 21(4), pp. 279–310.

Gautam, R., Sthapit, B.R. and Shrestha, P.K., 2006. Home gardens in Nepal: Proceeding of a workshop on enhancing the contribution of home garden to on: Farm management of plant genetic resources and to improve the livelihoods of Nepalese farmers: Lessons learned and policy implications, 6–7 August 2004. *Research and Development (LI–BIRD) PO Box,* 324.

Gautam, R., Suwal, R., Subedi, A., Shrestha, P.K. and Sthapit, B.R., 2004. Role of home garden in on: Farm agro: Biodiversity management and enhancing livelihoods of rural farmers of Nepal. *Paper presented in the Second National Workshop of in Situ Conservation Project,* 25–27 August, Nagarkot, Kathmandu.

Government of Nepal, 2018. *Labor migrant for employment: A status report for Nepal–2015/2016–2016/2017.* Available at: https://nepal.iom.int/sites/default/files/publication/Labour Migration_for_Employment-A_%20StatusReport_for_Nepal_201516201617_Eng.PDF

The Himalayan Times, 2019. Vegetables worth Rs 14bn imported in first six months, 24 March. Available at: https://thehimalayantimes.com/business/vegetables-worth-rs-14bn-imported-in-first-six-months/

The Indian Express, 2019. Nepal to import roses worth NPR 15 million from India for Valentine's day, 9 February. Available at: www.newindianexpress.com/world/2019/feb/09/nepal-to-import-roses-worth-npr-15-million-from-india-for-valentines-day-1936684.html

Joshi, B.K., Shrestha, R. and Karki, T.B., 2018. *Agrobiodiversity status, conservation approaches, good practices, neglected and underutilized species and future smart foods in Nepal.* Kathmandu, Nepal, National Workshop on Working Crop Groups.

Kattel, P., 2016. Socio: Economic importance of indigenous poultry in Nepal. *Poultry Fish Wildlife Science,* 4, p. 153.

Khanal, S., Khanal, D. and Kunwar, B., 2019. Assessing the structure and factors affecting agrobiodiversity of home garden at katahari rural municipality, Province 1, Nepal. *Journal of Agriculture and Environment,* 20, pp. 129–143.

Khanh, T.T., Ha, N.V., Phengsavanh, P., Horne, P. and Stür, W., 2006. March. The contribution of livestock systems to livelihood sustainability in the central highlands of Vietnam.

Towards Sustainable Livelihoods and Ecosystems in Mountainous Regions: Proceedings of an International Symposium. Chiang Mai, Thailand.

Ministry of Livestock Development (MOLD), 2017. *Annual report.* Available at: www.mold.gov.np/downloadfile/Tathaynka%20Pustika%20final%20for%20print_1510132862.pdf

Nair, P.K.R., 1991. State-of-the-art of agroforestry systems. *Forest Ecology and Management,* 45(1–4), pp. 5–29.

Nepali Sansar, 2018. *Culture: Nepal's Tihar calls for rs 110 million worth flower imports,* October 31. www.nepalisansar.com/news/nepal-tihar-calls-for-import-of-rs-110-million-worth-flowers/

Pulami, R. and Paudel, D., 2004. Contribution of home gardens to livelihoods of Nepalese farmers. In R. Gautam, B. Sthapit and P. Shrestha, eds. *Home gardens in Nepal.* Pokhara, Nepal: LI-BIRD, Bioversity International and SDC. Local Initiatives for Biodiversity, pp. 18–26.

Shrestha, P., Gautam, R., Rana, R.B. and Sthapit, B.R., 2002. Home gardens in Nepal: Status and scope for research and development. In J.W. Watson and P.B. Eyzaguirre, eds. *Home gardens and in situ conservation of plant genetic resources in farming systems, 17–19 July 2001.* Witzenhausen, Germany, Rome: IPGRI, pp. 105–124.

Sthapit, B., Gautam, R. and Eyzaguirre, P., 2006. The value of home gardens to small farmers. In R. Gautam, B.R. Sthapit and P.K. Shrestha, eds. *Home gardens in Nepal: Proceeding of a workshop on "enhancing the contribution of home garden to on: Farm management of plant genetic resources and to improve the livelihoods of Nepalese farmers: Lessons learned and policy implications," 6–7 August 2004.* Pokhara, Nepal: LI–BIRD, Bioversity International and SDC. Available at: www.bioversityinternational.org/fileadmin/_migrated/uploads/tx_news/Home_Gardens_in_Nepal_1166.pdf

Subedi, A., Suwal, R., Gautam, R., Sunwar, S. and Shrestha, P., 2006. Status and composition of plant genetic diversity in nepalese home gardens. In R. Gautam, B.R. Sthapit and P.K. Shrestha, eds. *Home gardens in Nepal: Proceeding of a workshop on "enhancing the contribution of home garden to on: Farm management of plant genetic resources and to improve the livelihoods of Nepalese farmers: Lessons learned and policy implications," 6–7 August 2004.* Pokhara, Nepal: LI–BIRD, Bioversity International and SDC. Available at: www.bioversityinternational.org/fileadmin/_migrated/uploads/tx_news/Home_Gardens_in_Nepal_1166.pdf

Sunwar, S., 2006. Home gardens in western Nepal: Opportunities and challenges for on. *Biodiversity and Conservation,* 15(13), pp. 4211–4238. DOI: 10.1007/s10531-005-3576-0

Talukder, A., Sapkota, G., Shrestha, S., de Pee, S. and Bloem, M., 2006. Homestead food production program in central and far: Western Nepal increases food and nutrition security: An overview of program achievements. In R. Gautam, B.R. Sthapit and P.K. Shrestha, eds. *Home gardens in Nepal: Proceeding of a workshop on "enhancing the contribution of home garden to on: Farm management of plant genetic resources and to improve the livelihoods of Nepalese farmers: Lessons learned and policy implications," 6–7 August 2004.* Pokhara, Nepal: LI–BIRD, Bioversity International and SDC. Available at: www.bioversityinternational.org/fileadmin/_migrated/uploads/tx_news/Home_Gardens_in_Nepal_1166.pdf

Tulachan, P.M. and Neupane, A., 1999. Livestock in mixed farming systems of the Hindu Kush-Himalayas. *Trends and Sustainability. ICIMOD, Kathmandu, Nepal, 116.* Available at: www.fao.org/3/x5862e/x5862e00.htm#TopOfPage

USAID, 2019. *Nepal.* Available at: www.usaid.gov/nepal/agriculture-and-food-security

Williamson, M. and Lyons, W., 2013. *Integrating livestock in the garden.* Available at: http://organicgrowersschool.org/wp-content/uploads/2013/12/41.-Livestock-in-the-garden.pdf

World Bank, 2019. *Nepal.* Available at: https://data.worldbank.org/country/nepal

10 Kitchen vegetable gardens for food and nutritional security of the poorest in rural India – experiences of the Aga Khan Rural Support Programme

Naveen K. Patidar, Sajan Prajapati and Kamlesh Panchole with inputs from field teams

Background: decreasing dietary diversity in rural India

Food and nutritional security are one of the key developmental challenges of rural India. All major surveys and research studies put India on the lower side when it comes to the food and nutrition security of its population. Some key facts on the nutritional status as per the National Family Health Survey of 2015–2016 conducted by the Ministry of Health and Family Welfare (2017) include:

- The mortality rate of children under five in rural India is close to 56 per 1,000 children, which is among highest in the world
- 38.3% rural children under five years of age in are underweight
- 26.7% of the rural women in India have below normal body mass index
- 59.5% rural children under five years of age are anemic
- 54.4% young women in rural India are anemic.

Non-availability of nutritious food along with poor preventive healthcare services are major reasons behind such a dismal food and nutritional security scenario in rural regions. In the last two to three decades, the cropping pattern in India has changed dramatically in favor of cash crops. There is also a decrease in dietary diversity due to overdependence on wheat and rice supplied by the Public Distribution System (PDS). In addition, farmers in India are incentivized to grow crops like wheat, paddy and sugarcane as government provide secured market facilities to farmers for these crops. This has led to a situation where farmers have slowly stopped growing crops like native vegetables, native fruits, pulses and millets which formed the foundation of nutritious diets for rural households and people. There is a continuous decrease in the area of millet cultivation over the last five decades. Millet is considered highly nutritious in nature. The International Crops Research Institute for the Semi-Arid Tropics (ICRISAT) on its website mentions that while the iron content of barnyard millet is 15.2 mg, and that of rice is 0.7 mg (Samuel, 2016). The calcium content of foxtail millet is 31 mg, and that of rice is 10 mg. While the percentage of nutrients varies with each variety of millet, in general they are richer in calcium, iron, beta carotene than rice and wheat. Millets are rich in dietary fiber, which is negligible in rice. With no gluten and a low glycemic index, the millet diet is ideal for those with celiac diseases and diabetes. A decrease in millet cultivation area particularly affected the food and nutrition security of smallholder farming families. According to AKRSP(I)'s own experience on ground and various group discussions done by its field teams, the consumption of vegetables in daily diet is found to be extremely limited in the case of the poorest sections of rural societies. They are mostly dependent on weekly village markets for vegetable purchase. The high cost of vegetables in the market discourages poor families from including vegetables in their food basket. According to a study named Enriching Dietary Diversity through Self-Provisioning Potential, Issues and Practices (2018), conducted by the Vikas Anvesh Foundation in India, nutritious food items from forests have almost vanished as most forests are highly degraded in nature. This has particularly impacted the food and nutrition security of tribal communities in forest areas. Consumption of fruits is almost absent in the current scenario as most of the forests are degraded and fruits in markets are costly for families with limited economic means.

One of the trends in Indian agriculture is also the increasing use of pesticides and herbicides which led to a decrease in floral diversity in farms as well as backyard lands of smallholder farmers. According to a policy report named *Pesticide Use in Indian Agriculture: Trends, Market Structure and Policy Issues* (2017), published by National Institute of Agricultural Economics and Policy Research (ICAR), both total as well as per hectare usage of pesticides in India show significant increase after the year 2009–2010. In the year 2014–2015, pesticide usage was 0.29 kg/ha (GCA), which is roughly 50% higher than the use in 2009–2010.

Previously, there were many leafy plants in and around farms which were consumed by rural people, but over time these plants have disappeared as a result of monocropping systems and overuse of chemicals in agriculture. All of these factors have led to decrease in dietary diversity for current generation of rural people in India in comparison to previous generations. This further leads to a decrease in the availability of vital macro- and micro-nutrients for rural people.

There have been attempts by various research agencies to ascertain the link between dietary diversity and nutrient deficiency. An assessment by the National Nutrition Monitoring Bureau (NNMB) presents an extensive discussion on the relationship between dietary diversity and levels of malnutrition with a key focus on micro-nutrient deficiency. Phansalkar and Kundu (2018, p. 3) state

> two studies in India specifically link dietary diversity with the prevalence of micro-nutrient deficiency and other forms of malnutrition. A report

of Indian National Science Academy laying down priorities for research and action on micronutrient deficiencies bemoans the reduction in dietary diversity. It states that cereal-pulse based diets in India are deficient in several micro-nutrients such as iron, calcium, Vitamin A, riboflavin and folic acid causing "hidden hunger" and that these deficiencies are amplified in families who have insufficient incomes to afford green leafy vegetables, fruit or animal-based foods. It goes on to record the erosion in areas producing millets resulting in near absence of nutritious grains from diets of the poor.

About the Aga Khan Rural Support Programme (India)

The Aga Khan Rural Support Programme (India) is a non-denominational, non-government development organization. AKRSP(I) works as a catalyst for the betterment of rural communities by providing direct developmental support to local communities. The organization is active in over 2,700 villages of Gujarat, Madhya Pradesh and Bihar and has impacted over 1.5 million people from marginalized sections of the society. Over 80% of the households supported by AKRSP(I)'s work belongs to marginalized communities such as tribals, dalits, minorities and other backward classes. Women constitute over 60% of direct beneficiaries and thus form a core group for program interventions.

AKRSP(I) has pioneered various participatory development approaches in the country over last three decades and it has won various national and international accolades. The backbone of AKRSP(I)'s work is empowerment of rural communities, particularly the underprivileged and women through collectivization as well as promotion of individual enterprises. Building self-reliant people's institutions for food and nutrition security, financial inclusion, livelihood enhancement and improved rural governance is the heart of the organization's approach. Currently, the organization has five major intervention areas of rural development: (1) food and nutrition security; (2) drinking water and sanitation; (3) economic inclusion; (4) civil society development; and (5) education and early childhood development.

AKRSP(I) and the kitchen gardens initiative

AKRSP(I) has been engaged in food security–related interventions since its inception primarily through agriculture productivity enhancement–related activities. In the initial years of its operations, the focus was on land development, irrigation development and agriculture extension–related activities. According to several evaluation reports (TTC, 2018; NRMC, 2016) of AKRSP(I)'s work, these initiatives led to a substantial increase in food crop production; however, most of this was around cereals. While there were some advances in increasing milk production in poor regions, it was found that smallholders were selling milk to cooperatives rather than consuming it at their household level. This situation required organizations to develop a strategy where issues related to dietary diversity are addressed rather than just focus on increased production of cereals and milk in villages. There were also issues related to the lack of land availability among a large proportion of households in rural areas. Over 40% households in Bihar and around 20% in Gujarat and Madhya Pradesh do not have any agriculture land, hence agriculture extension-related projects have a limited impact on this set of households.

Considering the aforementioned situation, AKRSP(I) started work on kitchen gardens which can address needs of a majority of the households in villages. A kitchen garden is simply a small patch of land in the back or front of a rural houses; it can be dedicated to the production of vegetables and fruits, primarily for consumption at the household level. Major objectives of promoting kitchen gardens are as follows:

- Generating awareness among rural households, particularly among women, on the importance of dietary diversity for better nutrition status of the family members
- Improving the knowledge and skills of rural women around practices of cultivating vegetables at the household level
- Empowering girls and women through opportunity to get an equitable share in the diverse sources of nutrition
- Promoting dietary diversity among rural population, particularly the most marginalized population of rural areas, through provision of vegetable cultivation and consumption at the household level
- Reducing expenditure incurred by poorest households on the purchase of vegetables from markets
- Institutionalizing the process of vegetable cultivation at the household level to make itself sustainable so that minimum external support is required once established.

With these objectives, AKRSP(I) designed its interventions in its core geographies which includes the northern region of Bihar, the southwest region of Madhya Pradesh and Saurashtra, and the southern region of Gujarat. These are some of the neediest regions in India. Bihar ranks among the poorest states in the country, while tribal communities living in Madhya Pradesh and South Gujarat are also among the most marginalized communities in the country. Almost all households in intervention villages are smallholder farmers owning less than 1 ha of land, with annual income per household of less than $700. A majority of the farmers in these regions grow paddy, cotton, soybean and wheat. Around 20% of the households with some access to irrigation services also grow vegetables, mostly for self-consumption. Millets have totally disappeared from most of these villages over time, while cultivation of legume crops is continuously on the decline.

Kitchen garden designs

Two major categories of kitchen gardens were designed by the organization for introduction and scale-up in rural areas. These are "backyard kitchen gardens" and "landless gardens." The first category is suitable for households having some backyard land available with them, while the second category is suitable for households that do not have land.

Backyard kitchen gardens

Backyard kitchen gardens are generally set up in the land available adjacent to the rural households (Photo 10.1). In reality this land can be on either side of the household location; both the back of the house or at the front of the house. Based on the land availability, types of vegetables and numbers of plants of each vegetable are decided jointly by the organization team and village women. A good mix includes both creeper varieties of vegetables and other vegetables. In some cases where households have some extra land available, provision is also made for some fruit trees which are source of rich nutrients.

An area of 5 m² to 10 m² is developed into backyard garden for vegetable cultivation. Fruit plants are also planted in either the backyard of houses or on the boundaries of the farms wherever possible. A combination of fruit plant species like papaya, guava, lemon, Indian gooseberry, vegetables like brinjal, tomato, coriander, green leafy vegetables, and spices are generally taken up by the households for cultivation. Table 10.1 presents a typically combination of fruits/trees and vegetables used by various geographies when creating a backyard kitchen garden.

Landless garden

Landless gardens are suitable for families that do not have much or any backyard land or they are located in degraded rocky places where cultivation is difficult. Vegetable cultivation in landless gardens takes place in sacks or bags (Photo 10.2). These are popularly known among beneficiaries as *Bori Baghicha*, which literally means "garden in bags." Only creeper vegetables can be grown in sacks filled with soils. Creeper vegetables like sponge guard, bitter guard, bottle guard and pumpkin takes place in sacks wherever soil and land availability is difficult in the backyard of houses. Typically, a landless garden involves four to five empty sacks/bags which are filled with a mixture of soil and compost. The seeds are sown one or two days after the filling of sacks/bags with soils. They are then kept in a dry place with proper drainage. Some support is provided for the creepers to climb the nearby walls and roof of houses.

Photo 10.1 Backyard kitchen garden in Bihar
Source: AKRSP(I).

Table 10.1 Fruits and vegetables used by various geographies

Region	Vegetables	Fruits/trees
Madhya Pradesh	Brinjal, tomato, coriander, green leafy vegetables, bitter guard, bottle guard	Papaya, guava, lemon, drumstick
Bihar	Brinjal, okra, bitter guard, chili, green leafy vegetables, green peas, cabbage, radish, tomato, coriander, bottle guard	Papaya, lemon
Gujarat	Okra, brinjal, tomato, coriander, green leafy vegetables, bitter guard, bottle guard	Mango

Source: Prepared by AKRSP(I).

Photo 10.2 A landless garden (*Bori-Baghicha*) in the Dangs district of Gujarat
Source: AKRSP(I).

Some basic points of consideration when deciding the model of a kitchen garden

There are some of the basic considerations made by the project teams when selecting the model of intervention and varieties of vegetables:

- Land availability – that is, whether the household has access to backyard land or not
- Soil quality – to decide suitable model of kitchen garden
- Locally adapted varieties are given priority
- Tradition and taste preferences of the community
- Variety – to ensure diversity of vegetables to be grown

- Availability of ample sunlight important to site selection
- Land quality – backyard land must be well drained for proper germination and growth of the plants
- Seasonality of growing kitchen garden; sometimes varieties needs to be changed according to the season of the year. Season also affects water availability for vegetables.

Process of implementation

AKRSP(I) always makes all food security interventions highly participatory in nature for earning community ownership and ensuring sustainability beyond project funds. The organization uses women self-help groups (SHGs) as the platform of implementing its kitchen vegetable garden programs. Generally, these groups are engaged in savings and credit operations, however if used well, these groups are the best platforms for the extension of food security programs including vegetable cultivation. Typically, SHGs have 10 to 20 members and they meet one or more times in a month. Over 100,000 women have been mobilized in SHGs by the organization over time. The AKRSP(I) team uses their meetings as an opportunity to introduce the kitchen vegetable garden concept by emphasizing its importance from several angles critical from a community perspective. Discussions are held around diets; the availability of vegetables at the household level; money spent by families in the market to purchase vegetables; the nutritional value of various vegetables; equity in vegetable consumption among family members, particularly from a gender angle; availability of resources for vegetable cultivation like seeds, water and soils; and availability of the market in case surplus is available with any family. Program teams use several communication tools like posters, flip charts and videos to make the discussion interactive and insightful. Once these discussions are done, leaders of the SHGs are asked to compile a list of names of households interested in putting up a kitchen vegetable garden and the preferred design of the vegetable garden, either in a backyard garden or a landless garden. SHG members themselves decide upon the types of vegetables they want to grow in their gardens. If the members are cultivating vegetables for the first time, then AKRSP(I) generally supports in procurement of seeds from markets and takes care of around 75% of the seed cost from various project grants while the balance is generally contributed by the participating households in the form of cash. Seed procurement happens in bulk for a cluster of villages having 20 to 40 villages. Procurement of seeds from market is also done in a participatory manner, hence some SHG leaders along with program teams visit nearby shops to assess quality and cost, and make decisions on where to procure seeds. Once the bulk procurement of seeds is done, seed kits are packed by SHG members themselves. Seed kits contain specified numbers of seeds of each of the selected varieties of vegetables. Packed seed kits are sent to villages for distribution, again through the SHG platform to the members who had shown interest in growing the vegetable gardens.

Photo 10.3 Vegetable mini-kits distributed to women SHGs in Khargone district in Madhya
     Pradesh

Source: AKRSP(I).

One important step is to prepare land in the backyard for kitchen gardens or
fill sacks with soil if the household chose to go for a landless garden. It involves
soil treatment using farmyard manure. Households are supported through
knowledge extension on preparation of good-quality farmyard manures. Some
households also go for vermicomposting, which is used as organic fertilizer
in these kitchen gardens. Another important practice is controlling pests and
insects in the kitchen gardens. Over the period of the last five years, AKRSP(I)
has extensively promoted the use of organic pest repellents instead of chemical
pesticides to ensure chemical-free vegetables are available for consumption at
the household level. Generally, these pest repellents are prepared through local
materials like neem leaf extract, cow urine and leaves of other locally available

trees. These local preparations are found to be extremely effective if families spray it on in regular intervals. Once seed mini-kits are received by the family and land or sacks are ready for sowing, families sow seed as per a standardized package of practices (Photo 10.3). Other operations like watering, weeding, fertilizer application and so forth are done per the needs of the kitchen garden. Regular knowledge extension and handholding support is provided by program teams on the ground to further support home gardeners.

Sustainability of operations

Once rural households are supported through project grants, AKRSP(I) ensures that communities continue kitchen vegetable gardens through their own resources from next season onwards. This process is institutionalized through village institutions, mostly the federations of women SHGs. Typically, 100 to 400 SHGs are federated at cluster levels into a apex institution of women. Leaders and volunteers of these institutions conduct the process of seed procurement through the collection of money from individual households and procure the seeds in bulk. All operations are done through these community institutions, including money collection, seed procurement, preparation of mini-kits and distribution of these mini-kits. Sometimes these institutions also get support from government departments in terms of free seeds for their members. Hence the involvement of community institutions, particularly women SHGs and their cluster level federations, is extremely important to ensure the sustainability and continuity of the interventions once project grant support is withdrawn from the villages.

Production and consumption

The types of plants, their numbers and yields vary from place to place. The illustration in Table 10.2 is from a location in Madhya Pradesh. It shows the range of vegetables available to the household on a weekly basis for four months during the monsoon season, which demonstrates the nutritional diversity that can be achieved through kitchen gardens.

Table 10.2 Kitchen garden production example from Madhya Pradesh

Name of vegetable	No. of plants per kitchen garden	Appx. production in one week
Sponge gourd	8–10	400–500 gm
Bitter gourd	8–10	400–500 gm
Bottle gourd	8–10	1,000–1,500 gm
Okra	15–20	400–500 gm
Brinjal	15–20	800–1,000 gm
Broad bean (*Ballar*)	5–6	300–400 gm
Cucumber	8–10	400–500 gm
Spinach (*Palak*)	30–40	400–500 gm

Source: Prepared by AKRSP(I).

Most of these vegetables from the kitchen gardens are consumed by the households practicing the cultivation. One kitchen garden is sufficient to meet the vegetable requirements of a typical family for around four months and households generally practice it twice in a year. In a majority of the cases people also share or exchange vegetables with neighboring households at no charge. In some places like Bihar, many of the participating households do sell their additional produce in the markets, which earn them around INR 1,000–2,000 from the entire kitchen garden production. In Madhya Pradesh, AKRSP(I) has also worked on community-based cooking events to ensure that cooking takes place in a process which do not deplete the nutritional values of the vegetables.

Coverage

Over time, AKRSP(I) has covered over 42,000 households through kitchen vegetable garden initiatives (Photo 10.4). These households are spread across 24 districts of three states in India. Most of the participating households are from marginalized communities including tribal, scheduled castes or religious minorities. As an illustration, the organization covered over 22,000 households in Bihar in the last five years, out of which 45% belonged to scheduled caste households and another 23% belonged to religious minority households. Similarly, in Madhya Pradesh coverage is around 8,000 households, and all of them belong to scheduled tribal communities.

Impact on target communities

Community interactions, various case studies and records of community institutions provide some clear insights on the importance of kitchen vegetable gardens for the communities. Most participating households reported a huge improvement in the consumption of vegetables in their meals. Among most marginalized households, vegetables were totally absent from diets prior to interventions. After the adoption of a kitchen garden, families are able to consume vegetables at least four to five days in a week in comparison to just once or twice in a week. This is a significant gain for such families and a notable nutritional enhancement. In addition, through household vegetable production, the family saves around INR 2,000–3,000 in a year.

The household's reliance on vegetables from the market and purchase have dramatically declined as a result of undertaking kitchen gardening. In around 20% of the cases, it is found that households sell the surplus vegetables in the market and earn around INR 1,000–2,000 through such sales. Interestingly, the whole process is driven by women, and it is constantly emphasized that women and girls must consume the vegetables produced in kitchen gardens. Several group discussions on the ground confirmed that there is huge improvement

Photo 10.4 Kitchen garden in Dhar district of Madhya Pradesh
Source: AKRSP(I).

within households with respect to equity between men and women when it comes to consumption of vegetables. It can be assumed that this might have helped in improving dietary diversity for children, adolescent girls and women in the families (Box 10.1).

Box 10.1 Case study of Harshaben's kitchen garden: secured food, nutrition and social bonding

A resident of Juvanpur village of Kalyanpur taluka in Devbhumi Dwarka district of Gujarat, Hanshaben Ahmedbhai Parmar is a member of a household below poverty line. She has four members in her family.

She and her husband work as wage laborers. The family barely makes ends meet. During one of the training sessions of the SHG meeting, conducted by AKRSP(I) in Juvanpur village, 53 women agreed to do kitchen garden in their backyard land. Hanshaben learned about the benefits of kitchen gardening and planted the kitchen garden in the backyard of her house. In July 2018, after receiving the kit consisting of seeds of four types of vegetables, she started sowing in the backyard of her house, toiling hard and following the instructions provided by AKRSP(I) field agents.

Harshaben took care of the plants, watered them at regular intervals and kept the garden healthy. By the third week of July, her efforts were rewarded. She planted seeds of bottle gourd (*Dudhi*) and bitter gourd (*Karela*), sponge gourd (*Galka*) and spinach (*Palak*). There was plenty of production of bottle gourd and spinach in her kitchen garden within two months of sowing the seeds. After retaining the adequate amount of vegetables for family consumption, Harshaben got approximately 4 kg per week as surplus from the kitchen garden.

Besides self-consumption, not buying any vegetables for home consumption helped her save INR 200 per week. Surplus vegetables generated helped increase the family income.

Also, she shares some bottle gourd and bitter gourd with her neighbors, which enables her to strengthen the social bonds with her community. As she explains:

> I definitely believe that there is an improvement in the nutrition status of my family, though it is not possible to measure the same at this stage. However, the diversification of the menu in my family and the regular intake of nutrient rich vegetables has improved the energy and efficiency levels of my family members.

Overall, the success of Harshaben's kitchen garden did not simply yield socio-economic benefits but also empowered her as a woman and a mother. Being the primary caretaker of her garden, she is now able to ensure nutritious food for her family.

"I used the savings and income generated from the vegetable garden to cater to other household needs like buying stationeries for my school going children," adds Harshaben with a smile.

Learnings

There is significant learning from AKRSP(I)'s experience with the kitchen garden initiative. Community and community-based institutions are extremely important for a sustainable initiative beyond project grants. If interventions are not thought through, then post-project continuation becomes a challenge and communities will discontinue kitchen gardening when the project teams withdraw from the villages. Women must be at the center of project programing as they drive the food requirements of the families in rural areas. If they are empowered with adequate knowledge around the importance of dietary diversity and are provided easy access to a variety of vegetable seeds, then success is certain. Affordable and ready to use vegetable mini-kits are one of the most important ingredients of any successful kitchen gardening initiative. Designs of the kitchen gardens according to community taste, local geography and traditions are also important to develop ownership by the community. Safe food is an important issue for households, hence cultivation of vegetables through farmyard manures and organic pest repellents is also gaining acceptance among the community members.

Way forward

AKRSP(I) is in process of making the food and nutrition security initiatives comprehensive in nature. Efforts are being made to include nutritious fruits as part of kitchen garden initiative; over 4,000 families have been covered in Bihar and Madhya Pradesh through this additional piece within kitchen garden initiative. Traditional millets are also being introduced in tribal regions so that there is complete access to nutrient-rich foods at the household level. Experiences of some other agencies in India suggest the importance of bringing back some of the wild varieties of fruits and vegetables into the diets, particularly for tribal communities. This can be looked at by AKRSP(I) in the future. Women's institutions will remain the backbone of the organization's kitchen garden initiative in the near future.

References

Ministry of Health and Family Welfare, 2017. *National family health survey of India-IV, 2015–16*. New Delhi, India: Government of India.

National Nutrition Monitoring Bureau, 2003. Prevalence of micro: Nutrient deficiencies. *NNMB Technical Report No. 22*. Hyderabad, India: National Nutrition Monitoring Bureau.

NR Management Consultants, 2016. *End term project evaluation, MGNREGA: From wages to sustainable development*. New Delhi, India: NRM Consultancy.

Phansalkar, S. and Kundu, S., 2018. *Enriching dietary diversity through self: Provisioning potential, issues and practices*. Pune, India: Vikas Anvesh Foundation.

Samuel, J., 2016. *Eat millets, pay less, stay healthier, save earth*. [Online] Available at: www.icrisat.org/eat-millets-pay-less-stay-healthier-save-earth/ [Accessed 26 November 2019].

Subash, S.P., Chand, P., Pavithra, S., Balaji, S.J. and Pal, S., 2017. *Pesticide use in Indian agriculture: Trends, market structure and policy issues*. Report number: 43, New Delhi, India: National Centre for Agricultural Economics and Policy Research.

TTC, 2018. *End line impact assessment of the ABF-AKRSP "Dangi Vikas" project on sustainable livelihoods for tribal communities in the Dangs region of Gujarat*. New Delhi, India: Think Through Consultancy.

11 Lessons learned and the way forward on home gardens

Karimbhai M. Maredia and Joseph Guenthner

Global experiences and research suggest that home gardens have enhanced food and nutritional security and livelihoods in communities worldwide, both in normal situations and post-conflict/post-disaster situations. Extensive review of the literature on home gardens from developing countries in Asia, Africa, and Latin America strongly support the positive contributions of home gardens for social and economic development (Galhena et al., 2013; IFAD, 2014).

Research confirms that home gardens fulfill social, economic and cultural needs while providing important ecosystem services. Although the size, structure, functions and contributions of home gardens vary from country to country, home gardens fit well with the broader agenda of agricultural research and development. They serve as a locally available source of fresh food products, especially vegetables, fruits, roots and tubers, herbs, medicinal plants, honey and animal-derived foods (milk, eggs, meat). The literature pinpoints that home gardens provide supplemental sources of food and need to be integrated into overall agricultural development and food security programs.

Lessons from chapters in this book

Chapter authors have researched home gardening in places around the globe. Each chapter provides many lessons. We briefly list some of them in the following bullets. From the many lessons and discoveries, we selected several from each chapter to give readers a synopsis of their work.

Lessons from Chapter 1 – Understanding the global practice of home gardening

- Home gardens have endured the test of time and were important parts of ancient Egyptian, Mayan, Greek, Roman and Persian societies
- Home gardens are small, close to residences, species diverse, accessible to poor people, and sources of family food, materials and income
- Differences between gardens and farms include species type, harvest frequency, location, cropping pattern, technology, input costs, skills and assistance

- Gardens provide access to proteins, vitamins and minerals that can help alleviate undernourishment, nutrient deficiency, impaired child development and child and maternal mortality
- Garden limiting factors include lack of inputs, time requirements, pests, soil quality, weather, water, knowledge, theft, marketing issues and cultural barriers.

Lessons from Chapter 2 – Home gardens for nutritional security of men, women and children

- In the drive to reduce hunger with starch-based rice, wheat and maize, the role of fruits and vegetables in balanced diets has not received adequate attention
- A review of 30 agricultural interventions found that most increased food production but did not improve nutrition or health, with one exception: home gardens
- Home gardeners can make efficient use of limited resources and can empower women to help their families and communities overcome malnutrition and sell excess produce to increase household income.

Lessons from Chapter 3 – Keeping it close to home: home gardens and biodiversity conservation

- In addition to food availability, home gardens help enhance livelihoods by producing medicinal plants, ornamentals, feed, fuel and building materials
- Some gardeners integrate livestock and fishponds into their plots, providing protein and income
- Many gardeners foster diversity by including different varieties, landraces and species, serving as reservoirs of biodiversity conservation.

Lessons from Chapter 4 – Gender and home gardens: toward food security and women's empowerment

- Researchers used stakeholder interviews and women focus groups to discover the role of women in food production in Tajikistan
- Gender bias has limited the roles of women on commercial farms, but women have "freedom to farm" on home gardens
- Women farmers and gardeners described a "triple burden" of working for family, household and field as being one the challenges they face.

Lessons from Chapter 5 – Home gardens for better health and nutrition in Mozambique

- A program in Cabo Delgado, Mozambique, is coupling nutrition education with demonstration gardens, including sweet potato for vitamin A, pumpkin for long storage, sesame for iron and *Moringa* (drumstick tree) for drought resistance

- Participants had consumed vegetables, animal proteins and legumes rarely if at all; the least food diversity was during the "hungry months" of January and February
- Barriers to vegetable consumption included (1) high cost, (2) destruction by animals and (3) the false perception that vegetables must be cooked in oil to be healthy.

Lessons from Chapter 6 – Home garden experiences in Costa Rica

- Two projects in Costa Rica were designed for small farmers and gardeners to help achieve food security with components including small livestock, forest trees, organic practices, horticultural crops, soil conservation, child-care and extension education
- Participants identified vegetable marketing problems – high transport costs, brokers, lack of markets, low and unstable prices and bad roads – and proposed organizing a vegetable collection center
- Fruit trees (avocado, citrus, macadamia and soursop) were distributed, enabling intercropping with shade-loving coffee plants and providing two sources of income.

Lessons from Chapter 7 – Bio-innovations toward sustainable agriculture: success stories from India

- The authors present four case studies designed to augment incomes of farming communities in three regions of India
- Seeking short-growing-season crops with profitable prices, growers in Case #1 who introduced medicinal, aromatic and spice crops are now profitably marketing dried herbs, essential oils and hydrosols
- Bringing back traditional crops to improve food security, participants in Case #2 found low yields but high nutrition value with traditional vegetable crops
- Case #3 promoted soil health, biotechnology, high-quality planting material, grower training and agri-based enterprise resulting in successful demonstration plantings of turmeric, potato and vermicomposting
- In a rice-dominant area, Case #4 involved demonstration plantings of tissue-cultured crops, with success stories regarding turmeric, banana and waste utilization.

Lessons from Chapter 8 – Home gardens as a resilience strategy for enhancing food security and livelihoods in post-crisis situations: a case study of Sri Lanka

- A home gardening initiative launched by the government of Sri Lanka was designed to enhance food security, nutrition security and incomes of people impacted by civil war; the initiative included access to land, credit,

inputs, markets, information and advisory services as well as incentives and rewards for innovative households
- After the conflict ended and when the reconciliation and resettlement process was underway, the government promoted home gardening and conducted best home garden contests, which stimulated innovation; award-winning gardens became demonstration gardens for education.

Lessons from Chapter 9 – Complementarity between the home gardening and livestock production systems in Nepal

- Many home gardeners in Nepal integrate livestock into home gardens which helps increase household income and enhance nutrition
- A traditional practice is to allow livestock grazing in gardens during the dry winter season for a few days or weeks, leaving manure and urine that increase plant health and productivity
- Livestock help improve food security during lean seasons, such as drought
- In a patriarchal society home gardens can empower women socially, economically and nutritionally.

Lessons from Chapter 10 – Kitchen vegetable gardens for food and nutritional security of the poorest in rural India – experiences of the Aga Khan Rural Support Programme

- A small-farmer program in India focused on food and nutrition security in poor households using two types of gardens: backyard kitchen gardens and landless gardens
- Backyard kitchen gardens include papaya, guava, lemon and gooseberry as well as vegetables such as brinjal, tomato, and green leafy vegetables. Landless gardens are in small or rocky spaces, using soil-filled bags to grow creepers – sponge gourd, bitter gourd, bottle gourd and pumpkins – that climb walls and roofs
- Successful projects include involvement of community organizations, women involvement, knowledge of dietary diversity, access to a variety of vegetable seeds, vegetable mini-kits, community ownership and food safety.

Key lessons from home gardens

Based on research conducted around the world, the following key lessons have been learned from the implementation of home gardens programs in developing countries.

1 Home gardens enhance family food availability and diversity of food products through domestic production, and in some cases through food

products purchased from the income generated from surplus sales of food produced in home gardens

2   Home gardens serve as a first step to feeding people through local food systems in post-conflict and post-disaster situations. More empirical evidence on the value and importance of home gardens in conflict and post-conflict situations needs to be researched and documented

3   Home gardens have empowered women and enhanced family livelihoods

4   Home gardens protect local biodiversity and local varieties of vegetables, fruits, herbs and medicinal plants, which have high nutritional value in terms of micronutrients and provide health benefits

5   Cost-benefit analysis of home gardening is needed to determine economic value and develop models that measure impacts on nutrition, gender issues and long-term sustainability

6   Sustainable mechanisms need to be developed for a steady supply and access to inputs and new technologies. This would include local linkages for technical support, advisory services, training and mentoring of home gardeners through local extension systems and agricultural universities. Modern communication and social media tools (smartphones, internet, etc.) should be included. Greater cooperation is needed among stakeholders (government, NGOs, universities, private sector, etc.) for providing support to home garden programs

7   Mechanisms need to be developed for sharing best practices, experience sharing and knowledge exchange among home gardeners and stakeholders that support home gardens programs

8   More support is needed for market information and market linkages and branding of food products grown/produced in home gardens.

Way forward to enhance the role and contributions of home gardens

The implementation of home gardens has raised questions and issues that need to be further examined and researched (Galhena, 2012). Some of the key areas of future research include the following:

1   What are the long-term impacts of home gardens on household food and nutritional security and economic growth in post-conflict situations?

2   How can livestock and poultry be better integrated into current home gardening systems?

3   What long-term role can home gardens play in terms of dietary diversification and family nutrition and mental health?

4   What long-term role can home gardens play in terms of women's empowerment and gender equity?

5   What role can home gardens play in terms of building local markets and value chain systems for economic growth?

6   Cost-benefit analysis to determine the economic value of home gardening

7   Integration of home gardens into national policies
8   Long-term sustainability of home gardens
9   What role can research, education, training and outreach programs play in terms of enhancing productivity, scalability and sustainability of home gardens?

References

Galhena, D.H., Freed, R., Maredia, K.M. and Mikunthan, G., 2013. Home gardens: A promising approach to enhance household food security and wellbeing. *Journal of Agriculture and Food Security*, 2(8).

Galhena, H.D., 2012. *Home gardens for improved food security and enhanced livelihoods in Northern Sri Lanka*. A Ph.D. Dissertation Submitted to Michigan State University, East Lansing, Michigan, USA.

IFAD, 2014. *Lessons learned: Integrated homestead food production*. Published by the International Fund for Agricultural Development (IFAD). Page 20.

Index

Note: **Boldface** page references indicate tables. *Italic* references indicate figures and boxed text.